Viral Proteases and Antiviral Protease Inhibitor Therapy

PROTEASES IN BIOLOGY AND DISEASE

SERIES EDITORS:
NIGEL M. HOOPER, *University of Leeds, Leeds, United Kingdom*
UWE LENDECKEL, *Ernst-Moritz-Arndt University, Greifswald, Germany*

For other titles published in this series, go to
www.springer.com/series/6382

Uwe Lendeckel • Nigel M. Hooper
Editors

Viral Proteases and Antiviral Protease Inhibitor Therapy

Proteases in Biology and Disease

Springer

Editors
Uwe Lendeckel
Institute of Medical Biochemistry
and Molecular Biology
Ernst-Moritz-Arndt University
D-17487 Greifswald
Germany

Nigel M. Hooper
Institute of Molecular and Cellular
Biology
Faculty of Biological Sciences
University of Leeds
Leeds
LS2 9JT
UK

ISBN 978-90-481-2347-6 e-ISBN 978-90-481-2348-3
DOI 10.1007/978-90-481-2348-3

Library of Congress Control Number: 2009928306

Printed on acid-free paper

Springer is part of Springer Science+Business Media (www.springer.com)

Preface

This, the eighth volume in the *Proteases in Biology and Disease* series, focuses on the role of proteases in virus function and their potential as anti-viral targets.

The respiratory illness 'severe acute respiratory syndrome' (SARS) was first reported in Asia in November 2002, and rapidly spread to several other countries across the world. The SARS coronavirus (SARS-CoV) causes SARS and a key step in the replication of the virus is the cleavage of the viral polyproteins by the SARS-CoV main protease. In the first chapter of this volume, Wei-Zhu Zhong and colleagues describe the functional importance of this protease in the viral life cycle, which is an attractive target for developing drugs against SARS, and outline a rational, structure-based approach to inhibitor design.

More than 30 million people worldwide suffer from infection by human immunodeficiency virus type 1 (HIV-1) and today there are several types of anti-HIV drugs that target the three enzymes, reverse transcriptase, protease and integrase. The aspartic protease encoded by HIV-1 is an important target for antiviral therapy for AIDS. In Chapter 2 Jozsef Tözsér and colleagues describe the basic properties of HIV-1 protease, its importance in structure-guided drug design for AIDS, and discuss recent developments in antiviral therapy based on targeting the HIV 1 protease and drug resistant mutants.

It is currently estimated that 2.2% of the world's population is infected with Hepatitis C virus (HCV) which can be transmitted mainly through intravenous drug use and contaminated blood products. Currently therapeutic options for HCV are limited and in Chapter 3 Philip Tedbury and Mark Harris describe the HCV protease which is one of the principle novel targets for new anti-HCV agents.

The focus of Chapter 4 by Marion Kaspari and Elke Bogner is human cytomegalovirus (HCMV), one of eight human herpesviruses that can cause life threatening diseases in newborns and immunocompromised patients. Like many viruses, HCMV has evolved strategies to redistribute host proteins and to take over host functions to promote viral replication. The host cell's ubiquitin-proteasome system (UPS) mediates degradation of misfolded proteins but is also required for specific processing events important for apoptosis, the cell cycle, protein sorting, etc. Since these processes interfere with viral replication, proteasome inhibitors are now in focus as new targets for antiviral therapy.

Human T-cell lymphotropic virus type 1 (HTLV-1), like its better known relative HIV-1, is a human single-stranded RNA retrovirus that infects 20–30 million people worldwide. In Chapter 5 Jeffrey-Tri Nguyen and Yoshiaki Kiso describe the HTLV-1 aspartic protease and its potential as an anti-viral target.

The family of picornaviruses includes a number of important human and animal pathogens such as poliovirus, hepatitis A virus, coxsackievirus, human rhinovirus and foot-and-mouth disease virus. Although the last 25 years have seen an enormous increase in our knowledge and understanding of the molecular biology and pathogenicity of these viruses, at present no anti-viral substances have been approved for clinical use against picornaviral infections. In Chapter 6 Tim Skern and colleagues begin by explaining the situations in which an anti-viral against a particular picornavirus would be advantageous and identify the possible proteolytic activities against which anti-viral substances can be directed.

From this volume we are sure that the reader will see the potential for proteases to be the targets for effective anti-virals and hope that this volume in the *Proteases in Biology and Disease* series will provoke further research in this important area and be a valuable source of information on viral proteases. Finally, we would like to thank all the authors for their scholarly contributions.

January 2009

Uwe Lendeckel
Nigel M. Hooper

Contents

Contributors

Elke Bogner
Charité University Hospital Berlin, Institute of Virology
Helmut-Ruska-Haus, Charitéplatz 1, 10117 Berlin, Germany

Kuo-Chen Chou
Gordon Life Science Institute, 13784 Torrey Del Mar Drive, San Diego
CA 92130, USA
College of Life Science and Technology, Shanghai Jiaotong University
800 Donglin Road, Shanghai, 200240, China
College of Life Science and Biotechnology, Guangxi University, Nanning,
Guangxi, 530004, China
Institute of Image Processing & Pattern Recognition, Shanghai Jiaotong
University, 800 Dongchuan Road, Shanghai, 200240, China

Qi-Shi Du
Gordon Life Science Institute, 13784 Torrey Del Mar Drive, San Diego
CA 92130, USA
College of Life Science and Biotechnology, Guangxi University
Nanning, Guangxi, 530004 China

Mark Harris
Institute of Molecular and Cellular Biology, Faculty of Biological Sciences,
University of Leeds, Leeds, LS2 9JT, UK

Marion Kaspari
Charité University Hospital Berlin, Institute of Virology, Helmut-Ruska-Haus,
Charitéplatz 1, 10117 Berlin, Germany

Yoshiaki Kiso
Department of Medicinal Chemistry, Center for Frontier Research in Medicinal
Science and Twentyfirst Century COE Program, Kyoto Pharmaceutical
University, Yamashina-ku, Kyoto 607-8412, Japan

David Neubauer
Max F. Perutz Laboratories, Medical University of Vienna
Dr. Bohr-Gasse 9/3, A-1030 Vienna, Austria

Jeffrey-Tri Nguyen
Department of Medicinal Chemistry, Center for Frontier Research in Medicinal Science and Twentyfirst Century COE Program, Kyoto Pharmaceutical University, Yamashina-ku, Kyoto 607-8412, Japan

Hong-Bin Shen
Gordon Life Science Institute, 13784 Torrey Del Mar Drive, San Diego
CA 92130, USA
Institute of Image Processing & Pattern Recognition, Shanghai Jiaotong University, 800 Dongchuan Road, Shanghai, 200240, China

Suzanne Sirois
Gordon Life Science Institute, 13784 Torrey Del Mar Drive, San Diego
CA 92130, USA
Université du Québec à Montréal (UQAM), Chemistry Department, C.P. 8888 Succursale Centre-Ville, Montréal, Québec, Canada, H3C 3P8

Tim Skern
Max F. Perutz Laboratories, Medical University of Vienna
Dr. Bohr-Gasse 9/3, A-1030 Vienna, Austria

Jutta Steinberger
Max F. Perutz Laboratories, Medical University of Vienna, Dr. Bohr-Gasse 9/3, A-1030 Vienna, Austria

Philip Tedbury
Institute of Molecular and Cellular Biology, Faculty of Biological Sciences, University of Leeds, Leeds, LS2 9JT, UK

Jozsef Tozser
Department of Biology, Molecular Basis of Disease Program, Georgia State University, 24 Peachtree Center Avenue, Atlanta, GA 30303, USA
Department of Biochemistry and Molecular Biology, Research Center for Molecular Medicine, Medical and Health Science Center, University of Debrecen, Debrecen, Egyetem tér 1, Life Science Building, Hungary

Irene T. Weber
Department of Biology, Molecular Basis of Disease Program, Georgia State University, 24 Peachtree Center Avenue, Atlanta, GA 30303, USA
Department of Chemistry, Molecular Basis of Disease Program, Georgia State University, Atlanta, GA 30303, USA

Dong-Qing Wei
Gordon Life Science Institute, 13784 Torrey Del Mar Drive,
San Diego CA 92130, USA
College of Life Science and Technology, Shanghai Jiaotong University
800 Donglin Road, Shanghai, 200240, China

Ying Zhang
Department of Chemistry, Molecular Basis of Disease Program,
Georgia State University, Atlanta, GA 30303, USA

Wei-Zhu Zhong
Gordon Life Science Institute, 13784 Torrey Del Mar Drive, San Diego
CA 92130, USA

Chapter 1
Study of Inhibitors Against SARS Coronavirus by Computational Approaches

Kuo-Chen Chou, Dong-Qing Wei, Qi-Shi Du, Suzanne Sirois, Hong-Bin Shen, and Wei-Zhu Zhong

Abstract Called by many as the biology's version of Swiss army knives, proteases cut long sequences of amino acids into fragments and regulate most physiological processes. They are vitally important in life cycle and have become a main target for drug design. This Chapter is focused on a special protease that plays a key role in replicating SARS (Severe Acute Respiratory Syndrome) coronavirus, the culprit of SARS disease. The progresses reported here are mainly from various computational approaches, such as structural bioinformatics, pharmacophore modelling,

K.-C. Chou (✉)
Gordon Life Science Institute, 13784 Torrey Del Mar Drive, San Diego, CA 92130, USA; College of Life Science and Technology, Shanghai Jiaotong University, 800 Donglin Road, Shanghai, 200240, China; College of Life Science and Biotechnology, Guangxi University, Nanning, Guangxi, 530004, China; Institute of Image Processing & Pattern Recognition, Shanghai Jiaotong University, 800 Dongchuan Road, Shanghai, 200240, China
e-mail: kcchou@gordonlifescience.org

D.-Q. Wei
Gordon Life Science Institute, 13784 Torrey Del Mar Drive, San Diego, CA 92130, USA; College of Life Science and Technology, Shanghai Jiaotong University, 800 Donglin Road, Shanghai, 200240, China

Q.-S. Du
Gordon Life Science Institute, 13784 Torrey Del Mar Drive, San Diego, CA 92130, USA; College of Life Science and Biotechnology, Guangxi University, Nanning, Guangxi, 530004, China

S. Sirois
Gordon Life Science Institute, 13784 Torrey Del Mar Drive, San Diego, CA 92130, USA; Université du Québec à Montréal (UQAM), Chemistry Department, C.P. 8888 Succursale Centre-Ville, Montréal, Québec, Canada, H3C 3P8

H.-B. Shen
Gordon Life Science Institute, 13784 Torrey Del Mar Drive, San Diego, CA 92130, USA; Institute of Image Processing & Pattern Recognition, Shanghai Jiaotong University, 800 Dongchuan Road, Shanghai, 200240, China

W.-Z. Zhong
Gordon Life Science Institute, 13784 Torrey Del Mar Drive, San Diego, CA 92130, USA

U. Lendeckel and N.M. Hooper (eds.), *Viral Proteases and Antiviral Protease Inhibitor Therapy*, Proteases in Biology and Disease 8,

molecular docking, and peptide-cleavage site prediction, among others. It is highlighted that the compounds $C_{28}H_{34}O_4N_7Cl$, $C_{21}H_{36}O_5N_6$ and $C_{21}H_{36}O_5N_6$, as well as KZ7088, a derivative of AG7088, might be the promising candidates for further investigation, and that the octapeptides ATLQAIAS and ATLQAENV, as well as AVLQSGFR, might be converted to effective inhibitors against the SARS protease. Meanwhile, how to modify these octapeptides based on the "distorted key" theory to make them become potent inhibitors is explicitly elucidated. Also, a brief introduction is given for how to use computer-generated graphs to rapidly diagnose SARS coronavirus. Finally, a step-by-step protocol guide is given on how to use ProtIdent, a web-server developed recently, to identify the proteases and their types based on their sequence information alone. ProtIdent is a very user-friendly bioinformatics tool that can provide desired information for both basic research and drug discovery in a timely manner. With the avalanche of protein sequences generated in the post-genomic age, it is particularly useful. ProtIdent is freely accessible to the public via the web-site at http://www.csbio.sjtu.edu.cn/bioinf/Protease/.

Keywords SARS • coronavirus proteinase • KZ7088 • AG7088 • binding pocket • octapeptide inhibitors • distorted key theory • ProtIdent web server

Abbreviations CoV: Coronavirus; M^{pro}: Main proteinase; SARS: Severe acute respiratory syndrome

1.1 Introduction

Proteases, also termed as proteinases, or peptidases, are enzymes that are essential for all stages of life cycle, such as conception, birth, growth, ageing, and death of all organisms. Functioning as biology's version of Swiss army knives, they cleave long sequences of amino acids into fragments, a process that is indispensable for the synthesis of all proteins, controlling protein composition, size, shape, turnover and ultimate destruction. Therefore, it can help us better understand the biology of life and the treatment of diseases to study proteases, their receptors and inhibitors.

A respiratory illness called "severe acute respiratory syndrome" (SARS) was first reported in Asia in November 2002. The illness spread to more than two dozen countries in North America, South America, Europe, and Asia within only a few months. Patients suffering from SARS usually begin having a high fever (>38°C or 100.4°F) with symptoms such as headache, malaise, chilly, rigor, diarrhoea, and body aches, followed by developing a dry (non-productive) cough and having trouble breathing that might be accompanied by or progress to hypoxia, a condition in which there is insufficient oxygen reaching body tissues. Most patients developed pneumonia with a fatality rate around 15%.

It is known that the culprit that causes SARS is the SARS coronavirus (sometimes shortened to SARS-CoV) (Peiris et al., 2003). It is also known that a key step for the replication of the culprit is cleaving the SARS-coronavirus polyproteins by a special protease, the so-called SARS coronavirus main protease (SARS CoV Mpro). The functional importance of the protease in the viral life cycle has made it an attractive target for developing drugs against SARS. To conduct the rational (or structure-based) drug design, a key step is to understand the binding interaction of SARS CoV Mpro with its ligands.

Progress in synthesis of novel test compounds for antiviral chemotherapy of SARS has been summarized by Kesel (Kesel, 2005), and that in drug discovery against SARS-CoV reported in a recent review (Wu et al., 2006). This chapter will focus on the progress mainly from the approaches of computer-aided drug discovery.

1.2 Binding Interactions

Based on the atomic coordinates of SARS-CoV Mpro (Anand et al., 2003), two binding models were developed (Chou et al., 2003). One is for the binding interaction of SARS-CoV Mpro with a compound called KZ7088 (Fig. 1.1a), and the other is for that with the octapeptide AVLQSGFR.

a KZ7088

b AG7088

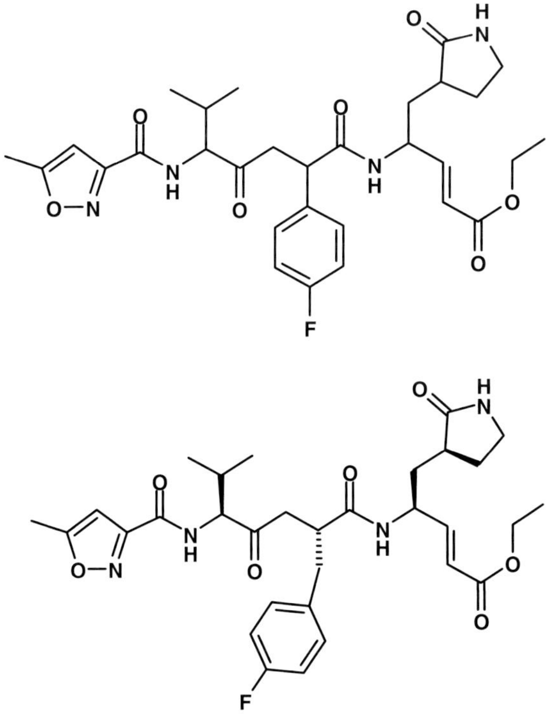

Fig. 1.1 Chemical structure of KZ7088: (**a**) a derivative of AG7088; (**b**) by removing –CH2 from its fluorophenylalanine side chain (Reproduced from Chou et al., 2003. With permission)

KZ7088 (Chou et al., 2003) is a derivative of AG7088 (Fig. 1.1b). The latter was developed by Pfizer Inc. and is currently in clinical trials for the treatment of rhinovirus, a pathogen that can cause the common cold. As shown in Fig. 1.1, AG7088 has a p-fluorophenylalanine side chain (p-fluorobenzyl), which is too long (or bulky) to fit into the binding pocket of SARS-CoV Mpro. By removing –CH2 from the side chain, the compound KZ7088 thus obtained could well fit into the binding pocket, as shown in Fig. 1.2. The constituents of the binding pocketare defined by those residues that have at least one heavy atom (i.e., other than hydrogen) with a distance ≤5 Å from a heavy atom of KZ7088. A similar bindingpocket was defined for ATP in the Cdk5-Nck5a*-ATP complex (Chou et al., 1999) and other studies (see, e.g., Zhou and Troy, 2003, 2005a, b; Chou, 2004b; Wei et al.,2006a; Wang et al., 2007a, b; Li et al., 2007a, b; Gu et al., 2009; Gong et al., 2009). The binding pocket of SARS-CoV Mpro for KZ7088 involved 23 residues, and the ligand was tethered to the enzyme by six hydrogen bonds, as detailed in (Chou et al., 2003). A series of follow-up discussions about the binding mechanism can be found in (Wu et al., 2006; Zhang and Yap, 2004; Samee, 2005; Clercq, 2006).

The binding interaction of SARS-CoV Mpro with the octapeptide AVLQSGFR is illustrated in Fig. 1.3, from which we can see that the octapeptide is tethered to Arg-40, His-41, Phe-185, Asp-187, and Gln-189 of SARS-CoV Mpro by six hydrogen bonds. The binding interaction mode had the important implications in stimulating rationally designing drugs for SARS therapy due to the following considerations: (1) The protease-susceptible sites in proteins usually extend to an octapeptide, as generally formulated by $P_4P_3P_2P_1P_{1'}P_{2'}P_{3'}P_{4'}$ with the scissile bond located between the subsites P_1 and $P_{1'}$, as generally expressed by P1↓P1′ (Schechter and Berger, 1967; Miller et al., 1989; Chou, 1993b). (2) The SARS coronavirus enzyme and several viral proteinases exhibit Gln↓ (Ser, Ala, Gly) specificity (Anand et al., 2003). (3) According to the "lock-and-key" mechanism in enzymology, the octapeptide cleavable by the SARS protease must have a good fit for binding to the active site. However, such a peptide, after a modification of its scissile bond with some simple routine procedure, will completely lose its cleavability but it can still bind to the active site. Actually, the molecule thus modified can be compared to a "distorted key" (Chou, 1996), which can be inserted into a lock but can neither open the lock nor be pulled out from it, spontaneously becoming an ideal competitive inhibitor against the SARS proteinase.

Stimulated by the above binding interaction modes, a series of follow-up studies were conducted, as described below.

1.3 Narrow down the Compounds-Searching Scope by Pharmacophore Approach

As mentioned above, both KZ7088 and the octapeptide AVLQSGFR are tethered to SARS-CoV Mpro by six hydrogen bonds. It is instructive to point out that analyzing the hydrogen bonding interactions of a receptor with its ligand often provided

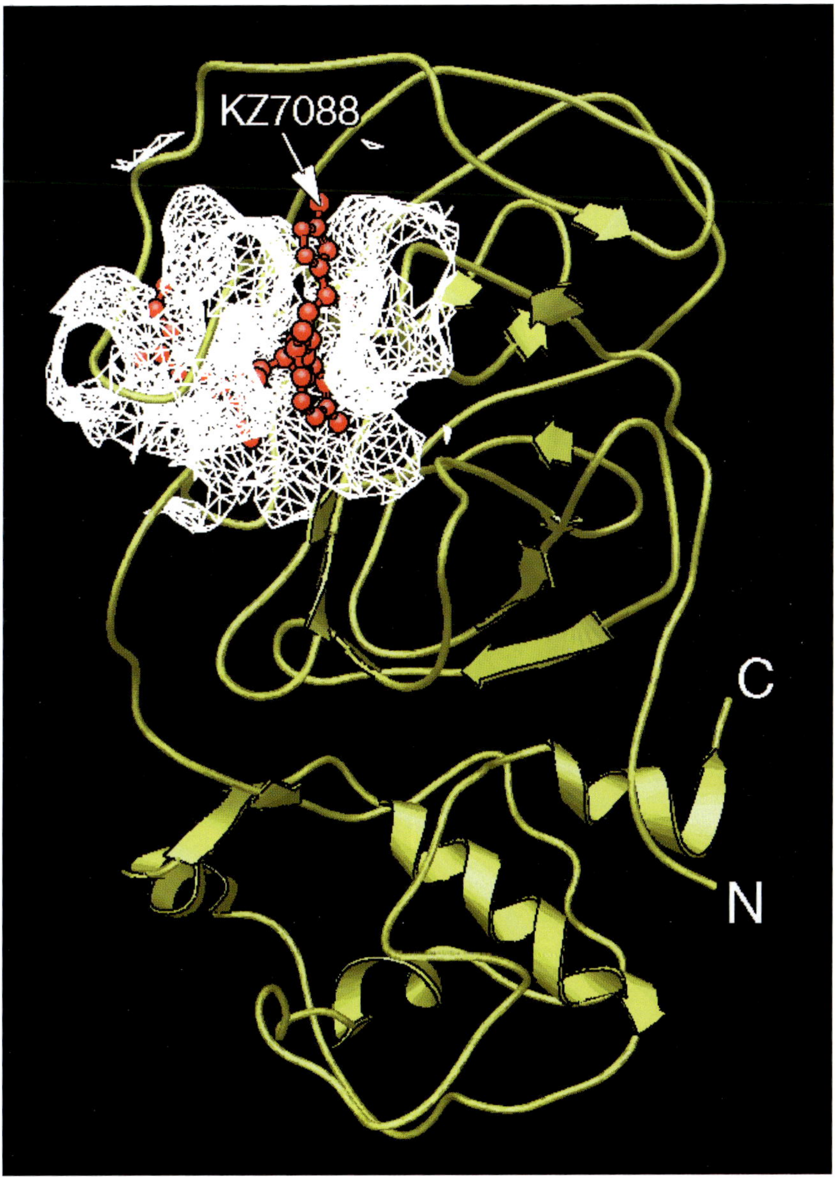

Fig. 1.2 An overall view of the complex obtained by docking KZ7088 to SARS-coronavirus main proteinase, where the SARS enzyme is colored yellow (in ribbon drawing), the KZ7088 colored red (in ball-and-stick drawing), and the white mesh represents the binding pocket for KZ7088 (Reproduced from Chou et al., 2003. With permission)

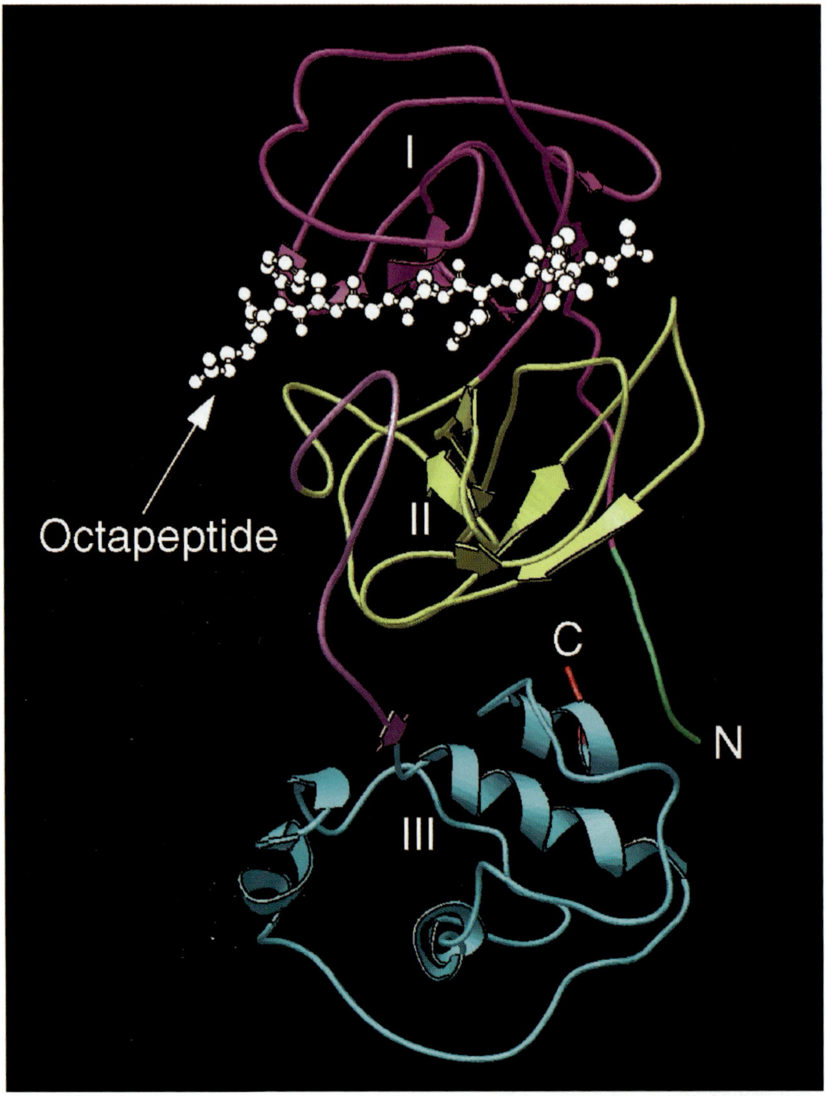

Fig. 1.3 An overall view of the complex obtained by docking the octapeptide AVLQSGFR to SARS-coronavirus main proteinase, where the octapeptide is coloured white (in ball-and-stick drawing), and the SARS enzyme is in ribbon drawing. As shown in the figure, the octapeptide is inserted into the cleft between domains I and II as expected (Reproduced from Chou et al., 2003. With permission)

useful insights for drug development (see, e.g., Chou, 2004a, b, 2005a, b; Chou & Howe, 2002; Wei et al., 2005, 2006a, b, 2007; Wang et al., 2007a, b, c, d, 2008; Li Y et al., 2007; Chou et al., 1997, 2000; Du et al., 2004, 2005a, 2007a, b; Zhang et al., 2006; Gao et al., 2007; Li L et al., 2007; Zheng et al., 2007; Guo et al., 2008; Schnell et al., 2008).

Based on the hydrogen bonding interactions of SARS-CoV Mpro with KZ7088 (Chou et al., 2003), a template of pharmacophore points of KZ7088 was generated (Sirois et al., 2004) as shown in Fig. 1.4, where the four hydrogen bond acceptors are coloured in cyan, the two hydrogen bond donors in purple, the aromatic group in green, and the volume constrains in grey. Based on the pharmacophore template, the virtual screening search operation was performed by Sirois and co-workers (Sirois et al., 2004) for the commercially and academically available compounds. It was found by pharmacophore searches that, of the 3.6 millions of compounds screened, 0.07% were with the score satisfying five of the six pharmacophore points as defined by the four hydrogen bond acceptors and two hydrogen bond donors (Fig. 1.4). Moreover, each of the hit compounds was further evaluated for its drugability according to a novel score function, an equation formulated by 13 metrics that took into account physical (Lipinski, 2000), chemical and structural properties (Xu and Stevenson, 2000), as well the presence of undesirable functional groups (Oprea, 2000; Rishton, 1997). The selection of these properties was based on the work proposed by Baurin (Baurin, 2002). After such a drugability-evaluating procedure, it was found that 17% of the compounds thus retrieved from the first hit screen had a perfect score of 1.0, 23% with one violation of drugable rule, 13% with two violations, and 47% with more than two violations. If the criterion for drugability was set at a maximum allowance of two violations, we obtained that only $0.07 \times (0.17 + 0.23 + 0.13) = 0.037$

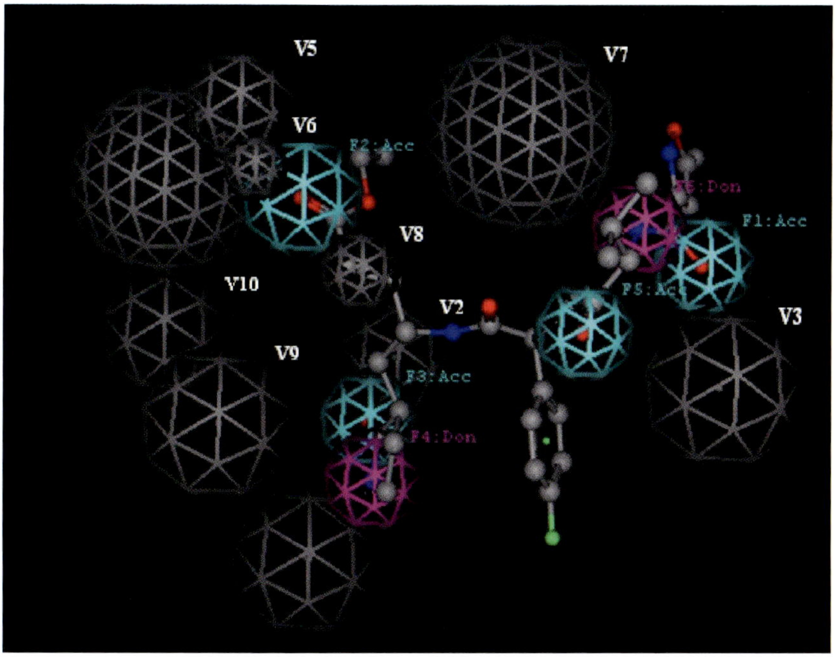

Fig. 1.4 A template of the pharmacophore points of KZ7088 derived from its binding interaction with SARS-CoV M^{pro} (Chou et al., 2003). The four H-b donors are colored in cyan, the two H-b donors in purple, the aromatic group in green, and the volume constrains in grey (Reproduced from Sirois et al., 2004. With permission)

of the total compounds screened are worthy of further tests for their activities. These findings would significantly narrow down the search scope for potential compounds, saving substantial time and money. Finally, the featured templates derived from the pharmacophore study would also be very useful for guiding the design and synthesis of effective drugs for SARS therapy.

It is instructive to point out that, instead of choosing the hydrogen bond donors and acceptors for pharmacophore modelling as done in (Sirois et al., 2004), one can also choose some other features for modelling, such as aromatic ring, hydrophobic aromatic, hydrophobic aliphatic, positive charge, negative charge, hydrogen bond acceptor lipid, positive ionizable, and negative ionizable.

1.4 Winnow for Anti-SARS Drug Candidates by Molecular Docking Operation

From the 2,589 unique 3D hits with volume constrains found by Sirois et al. (2004), 1,386 drugable compounds were collected by Wei and co-workers (Wei et al., 2006) that both match the 3D (dimensional) KZ7088 pharmacophore points and satisfy the drugable rules with a scoring value ≥0.8. It was found via similarity search and auto docking operation that only three compounds, i.e., $C_{28}H_{34}O_4N_7Cl$, $C_{21}H_{36}O_5N_6$, and $C_{21}H_{36}O_5N_6$ (Wei et al., 2006), were the most promising candidates worthy of further investigation.

Meanwhile, a thermodynamic analysis (Wei et al., 2006) indicated that the hydrogen bonds make an important contribution to the interactions of SARS-CoV Mpro with these compounds, which is quite consistent with the case of KZ7088 (Chou et al., 2003). In addition to hydrogen bonds, there are some other contributions to the interactions, such as desolvation, hydrophobic interaction, and electrostatic interaction. The hydrogen bonds may play the role as an “anchor”, determining the spatial position of the compounds in the binding pocket, thus facilitating the hydrophobic interactions and electrostatic interactions. Therefore, in conducting rational drug design it is critic to identify the hydrophobic groups of the ligand and receptor and make sure that they are facing to each other upon binding. Also, it was observed by examining the Connolly surface (Connolly, 1993) of the complex of SARS-CoV Mpro with each of the three compounds that the steric complementarity is responsible for the micro-mechanical interlock at the molecular level. In other words, it is the steric complementarity between the ligand and the collagen receptor site that plays the role of the primary driving force for mechanical interlocking.

The main features shared by the aforementioned three potential inhibitors as well as the information of the involved side chains of SARS-CoV Mpro may provide useful insights for the development of potent inhibitors against the SARS enzyme.

Recently, Wang et al. performed a virtual screening for finding natural inhibitor against cathepsin-L for SARS therapy (Wang et al., 2007d).

1.5 Distorted-Key Theory and Peptide Inhibitors

Owing to their inability to readily crossing through membrane barriers such as the intestinal and blood-brain barriers, developing peptide drugs is limited by their poor metabolic stability and low bioavailability. However, peptide drugs are of low toxicity to human body than organic compounds, and hence systematic chemical modification strategies that convert peptides into drugs are an attractive research topic in current medicinal chemistry (Adessi and Soto, 2002). Some efforts have been made in an attempt to develop peptide inhibitors against the SARS-CoV M^{pro} (Chou et al., 2003; Du et al., 2004, 2005a, b, 2007b; Zhang et al., 2006). The development of peptide inhibitors against proteases was based on the "distorted key theory" (Chou, 1996), as can be illustrated as follows.

According to the "lock-and-key" mechanism in enzymology, a protease-cleavable peptide must satisfy the substrate specificity, i.e., a good match for binding to the active site. Here, the phrase of "good match" should be understood in a broad sense rather than a narrow geometric sense; i.e., it means a favourable chemical-group-disposition for the binding of a substrate to the active site of an enzyme and the catalytic reaction thereof. However, such a peptide, after a modification of its scissile bond with some chemical procedure, will completely lose its cleavability but it can still bind to the active site of an enzyme. Actually, the molecule thus modified can be compared to a "distorted key", which can be inserted into a lock but can neither open the lock nor be pulled out from it. That is why a molecule modified from a cleavable peptide can spontaneously become a competitive inhibitor against the enzyme. An illustration about using the concept of "distorted key" to find peptide inhibitor for the SARS enzyme is given in Fig. 1.5, where panel (a) shows an effective binding of a cleavable peptide to the active site of SARS CoV M^{pro}, while panel (b) shows that the peptide has become a non-cleavable one after its scissile bond is modified although it can still bind to the active site. Such a modified peptide, or "distorted key", will automatically become an inhibitor candidate against SARS CoV M^{pro}. It is instructive to point out that, even for the case of developing non-peptide inhibitors, the knowledge thus acquired can also provide useful insights for the key binding groups, proper microenvironment, fitting conformation, among many other subtle requirements. Therefore, it will significantly reduce the search scope and expedite the process for finding the desired peptide inhibitors if a computational method is available by which one can discern what kind of peptides can be cleaved by SARS CoV M^{pro}. That is why during the process of developing inhibitors against HIV protease many algorithms were developed (Chou, 1993a, b; Poorman et al., 1991; Chou et al., 1993, 1996; Zhang and Chou, 1993; Thompson et al., 1995; Cai and Chou, 1998; You et al., 2005; Liang and Li, 2007; Rognvaldsson et al., 2007) for predicting the cleavable peptides by HIV protease. Very recently, a freely accessible web-server called HIVcleave (Shen and Chou, 2008) was also established for identifying the cleavage sites of proteins by HIV protease.

Stimulated by the distorted key theory for finding the peptide inhibitors against HIV protease, the similar bioinformatics tools were also developed (Gao et al., 2003; Chen et al., 2003) for finding the peptide inhibitors against SARS-CoV M^{pro}.

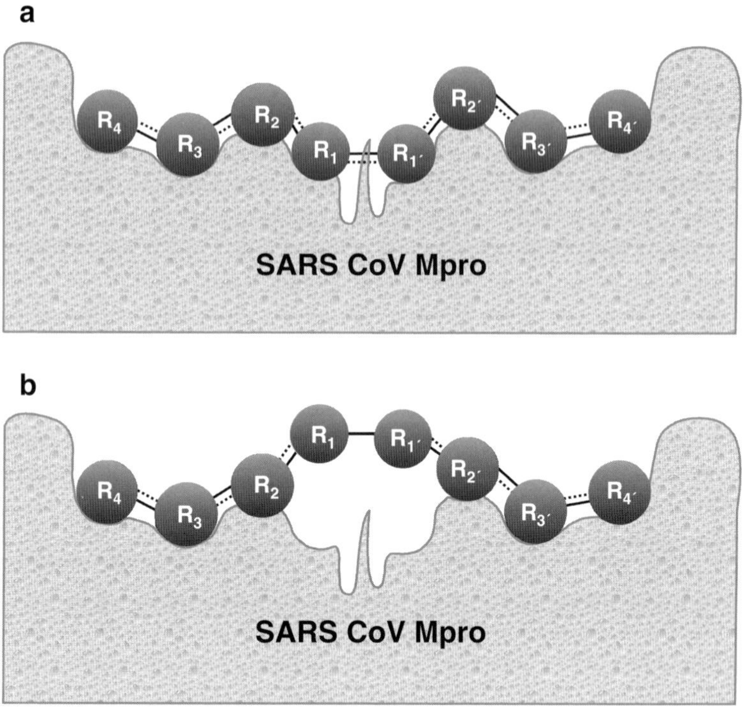

Fig. 1.5 Schematic illustration to show (**a**) a cleavable octapeptide is chemically effectively bound to the active site of SARS CoV Mpro, and (**b**) although still bound to the active site, the peptide has lost its cleavability after its scissile bond is modified to a strong hybrid peptide bond. The eight residues of the peptide is sequentially symbolized by R_4, R_3, R_2, R_1, $R_{1'}$, $R_{2'}$, $R_{3'}$, and $R_{4'}$, and their counter parts by S_4, S_3, S_2, S_1, $S_{1'}$, $S_{2'}$, $S_{3'}$, and $S_{4'}$, respectively (Adapted from Chou, 1996. With permission)

Du and co-worker (Du et al., 2005a) used these tools to search 36 SARS coronaviruses and found 396 cleavage sites, from which 11 cleavable octapeptides were extracted. In order to find the physical and chemical properties of the amino acids from positions R_4 to $R_{4'}$, Du et al. aligned the 11 cleavable peptides using the flexible molecular alignment technique (Du et al., 2007b). The results are given in Fig. 1.6, where panel (a) shows the lipophilicity and hydrophilicity of amino acid side chains of the 11 cleavable peptides, and panel (b) shows the pharmacophore types of atoms in the relevant amino acid side chains. As shown in the figure, residues at R_2 and R_4 are the most lipophilic; residues at $R_{1'}$ and $R_{3'}$ are partially lipophilic residues. On the other hand, residues at R_1 are the most hydrophilic, and residues at R_3 and R_4 are partially hydrophilic residues. The residues at $R_{2'}$ could be lipophilic or hydrophilic. The size of side chain of $R_{1'}$ is very small, and the volumes of side chains of residues $R_{3'}$, R_2 and $R_{3'}$ are larger than the others. Aromatic groups are on the positions R_2 and $R_{3'}$. Subsequently, the docking operations were performed for the 11 cleavable peptides with respect to SARS CoV M^{pro}. It was found that of the 11 cleavable peptides, NH_2-ATLQ↓AENV-COOH had the highest affinity with SARS CoV M^{pro}.

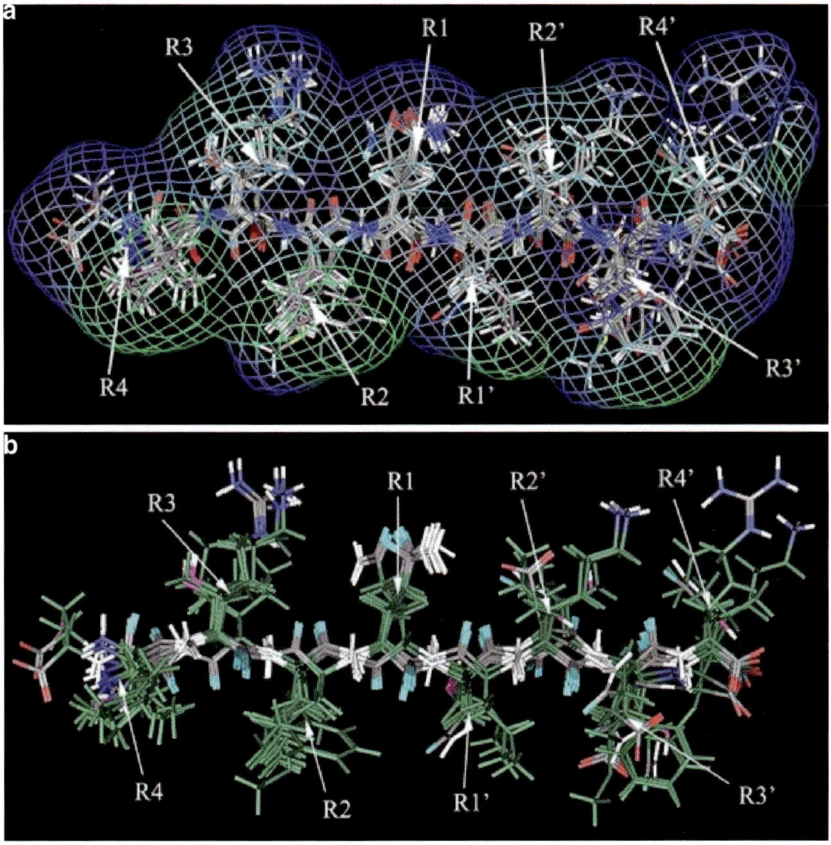

Fig. 1.6 The alignment of the 11 cleavable peptides. (**a**) Hydrophilic and hydrophobic surfaces of peptides. The green is for lipophilic regions and the blue is for hydrophilic regions. (**b**) Atomic pharmacophore types of peptides. Atoms are colored according to their pharmacophore type, donor is white, acceptor is cyan, acceptor and donor is magenta, cation or resonance cation is light blue, anion or resonance anion is red, hydrophobe is dark green, and gray is for other atoms

The binding interaction of the enzyme with NH_2-ATLQ↓AENV-COOH obtained by the docking operation is shown in Fig. 1.7.

Although a cleavable peptide can fittingly bind to the active site of a protease, it is not an inhibitor to the enzyme because it has a scissile peptide bond that will be eventually cleaved. However, a cleavable peptide can become an effective inhibitor after some proper chemical modification, as elaborated in Chou, 1996. Converting the cleavable peptides to stable and bio-available drugs is a hot research topic in medicinal chemistry (Chou, 1996, 2004; Du et al., 2004). Some possible chemical modification methods in this regard were analyzed (Du et al., 2007b). According to the distorted key theory (Chou, 1996), a cleavable oligopeptide can automatically become a competitive inhibitor if its scissile peptide bond between R_1 and $R_{1'}$ is replaced by a strong "hybrid peptide bond" (cf. Fig. 1.5). For the case of NH_2-ATLQ↓AIAS-COOH, the position R_1 and $R_{1'}$ is occupied by amino acids Gln and

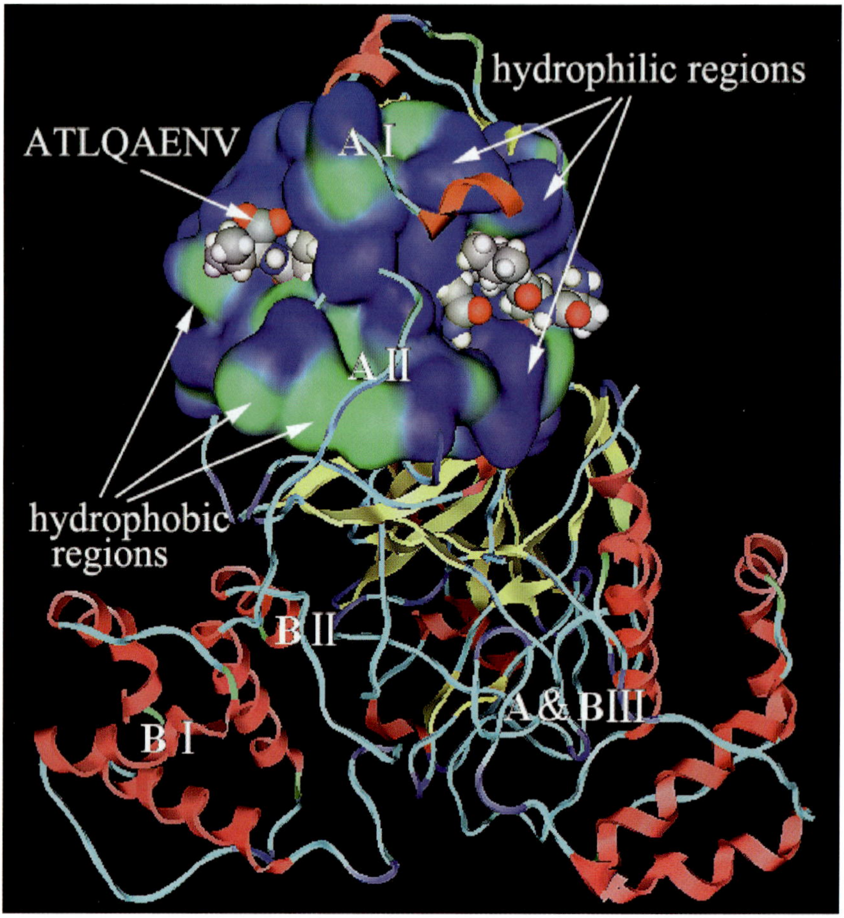

Fig. 1.7 The binding interaction obtained by docking the octapeptide ATLQAENV to SARS CoV M^{pro}. The green is for hydrophobic regions and the blue is for hydrophilic regions. The cleavable peptide P9 is shown as space filling model. The bioactive pocket is the cleavage cavity between domains I and II of protomer A

Ala. The peptide bond CO $\simeq$ NH between Gln and Ala is considered as a pseudo conjugate π-bond (here we use CO $\simeq$ NH to emphasize the conjugate π-bond), which is easily hydrolyzed by SARS CoV M^{pro}. If the group CO is replaced by CH_2, or NH is replaced by CH_2, the conjugate π-bond will become a strong single bond, and hence the hydrolyzation will become very difficult (Du et al., 2004; Szelke et al., 1982; Venkatesan and Kim, 2002). In general, the methylamino isostere is prepared by reduction of the amide bond with different reducing agents or by reductive amination of an amino aldehyde (Benkirane et al., 1996). The isostere has been incorporated into the ras farnesyl-protein transferase inhibitor that is a potential anticancer drug (Graham et al., 1994). Pseudo peptide analogs containing reduced amide isostere are also shown to inhibit C_α-N-C cleavage in the gag region of HIV-1 protease (Lowther et al., 1995). It was also observed by quantum chemistry calculations (Du et al., 2007b) that after CO was replaced by CH_2, the charge of the carbon atom in CH_2 turned to negative (–0.6572) from original a positive value (0.8612) in CO group, hence the nucleophilic attack by OH^- became impossible. Furthermore, the negative charge of N(NH) on the Ser side decreased to –0.2892 from –0.8190, and hence the electrophilic attack by H^+ became more difficult as well.

On the other hand, the experiments by Gan et al. (Gan et al., 2006) indicated that the octapeptide AVLQSGFR proposed originally in (Chou et al., 2003) was indeed cleavable by SARS-CoV M^{pro} with a high bioactivity. Based on the crystal structure of SARS-CoV M^{pro} (Yang et al., 2003), the cleavage mechanism of the SARS-CoV M^{pro} on the octapeptide NH_2-AVLQ↓SGFR-COOH (Chou et al., 2003) was investigated using molecular mechanics (MM) and quantum mechanics (QM) (Du et al., 2005b). It has been observed that the catalytic dyad (His-41/Cys-145) site between domain I and II of the protease [4,5] attracts π-electron density from the peptide bond Gln-Ser, increasing the positive charge on C(CO) of Gln and the negative charge on N(NH) of Ser, so as to weaken the Gln-Ser peptide bond. The catalytic functional group is the imidazole group of His-41 and the S in Cys-145. $N_{\delta 1}$ on imidazole ring plays the acid-base catalytic role. It has also been found that the chemical bond between Gln and Ser will become much stronger and no longer cleavable by the SARS enzyme after either changing the carbonyl group CO of Gln to CH_2 or CF_2, or changing the NH of Ser to CH_2 or CF_2. According to the "distorted key theory" (Chou, 1996), the octapeptide thus modified might become an effective inhibitor against SARS-CoV M^{pro} or a potential drug candidate for SARS therapeutics.

1.6 Using Graphic Approaches to Detect SARS-Coronavirus

By analyzing the two-dimensional cellular automata images (Wolfram, 1984, 2002) for 96 SARS coronavirus and 25 non-SARS coronavirus, Wang et al. (Wang et al., 2005) found a remarkable distinction between the SARS and non-SARS coronal viruses: all the images of the former were with a V-shaped cross-lines pattern (Fig. 1.8a) but all the images of the latter are with a parallel slash-lines pattern (Fig. 1.8b). They concluded that the V-shaped cross-lines image pattern might be deemed as a fingerprint for SARS coronal viruses for rapidly diagnosing SARS coronavirus.

Subsequently, Gao and co-workers (Gao et al., 2006) developed another graphic method for visualizing the characteristic of SARS-CoV. It has the merits of clear visibility and reliability that can serve as a complementary clinical tool for detecting SARS-CoV, particularly for the cases where the results obtained by the conventional methods are uncertain or conflicted with each other.

1.7 Identification of Proteases and Their Types

Owing to their indispensable role in life cycle, proteases have become an important target for drug discovery. Different types of proteases have different action mechanisms and biological processes. With the avalanche of new protein sequences generated

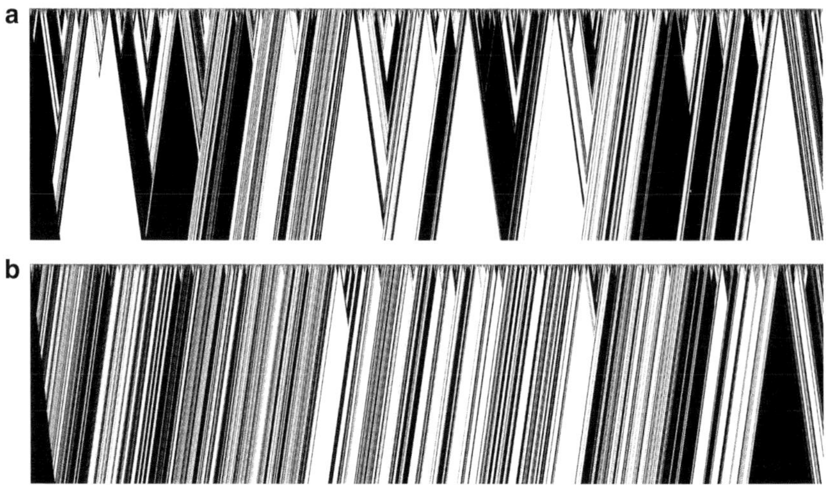

Fig. 1.8 The cellular automata images obtained by applying the modified Rule 184 (Wolfram, 1984, 2002) on (**a**) SARS coronavirus and (**b**) non-SARS coronavirus (Reproduced from Wang et al., 2005. With permission)

in the post-genomic age, it is highly desired to develop an automated method that can timely identify proteases and their types according to their sequence information alone.

Recently, a web server called ProtIdent (Protease Identifier) has been established at http://www.csbio.sjtu.edu.cn/bioinf/Protease/. ProtIdent is a two-layer predictor. The first layer is for identifying a query protein as protease or non-protease. If it is a protease, the process will automatically go to the second layer to further identify it among the following six functional types: (1) aspartic, (2) cysteine, (3) glutamic, (4) metallo, (5) serine, and (6) threonine. The success expectancy with ProtIdent in identifying a protein as protease or non-protease was about 92%, and that in identifying the protease type was about 96%.

As a user-friendly web server, ProtIdent is freely accessible to the public. Below, let us provide a step-by-step guide on how to use ProtIdent to get the desired results.

Step 1. Open the web page http://www.csbio.sjtu.edu.cn/bioinf/Protease and you will see the top page of the ProtIdent predictor on your computer screen, as shown in Fig. 1.9. Click on the Read Me to see a brief introduction about ProtIden predictor and caveat in using it.

Step 2. Either type or copy and paste the query protein sequence into the input box at the center of Fig. 1.9. The input sequence should be in the single-letter amino acid code, as shown by clicking on the Example button right above the input box. For instance, if your input is the 306 amino acids of SARS coronavirus main protease unit taken from the grey-boxed and underlined segment in **Box 1.1**, the input screen should look like the screenshot of Fig. 1.10.

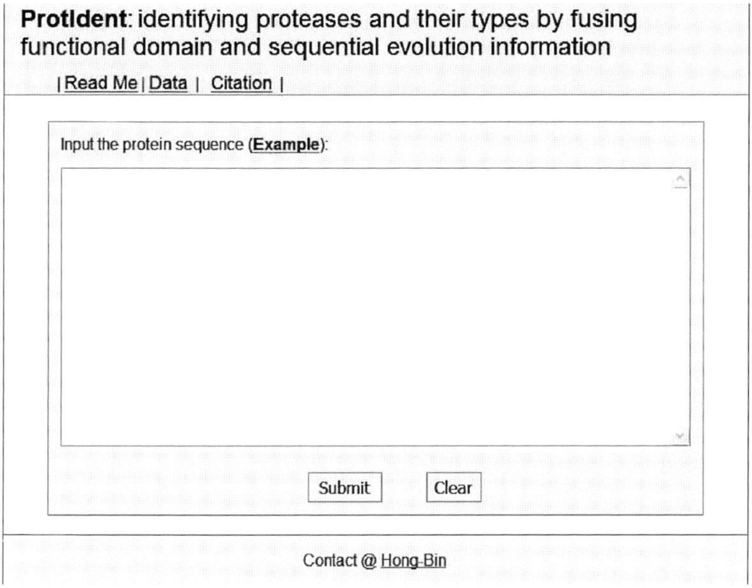

Fig. 1.9 A screenshot to show the top page of the web server ProtIndent at http://www.csbio.sjtu.edu.cn/bioinf/Protease

Box 1.1 The full-length amino acid sequence of SARS coronavirus protease (SW: P59641) where the grey-boxed and underlined segment (from res.3241 to res.3546) is the main protease

MESLVLGVNEKTHVQLSLPVLQVRDVLVRGFGDSVEEALSEAREHLKNGTCGLVELEKGV
LPQLEQPYVFIKRSDALSTNHGHKVVELVAEMDGIQYGRSGITLGVLVPHVGETPIAYRN
VLLRKNGNKGAGGHSYGIDLKSYDLGDELGTDPIEDYEQNWNTKHGSGALRELTRELNGG
AVTRYVDNNFCGPDGYPLDCIKDFLARAGKSMCTLSEQLDYIESKRGVYCCRDHEHEIAW
FTERSDKSYEHQTPFEIKSAKKFDTFKGECPKFVFPLNSKVKVIQPRVEKKKTEGFMGRI
RSVYPVASPQECNNMHLSTLMKCNHCDEVSWQTCDFLKATCEHCGTENLVIEGPTTCGYL
PTNAVVKMPCPACQDPEIGPEHSVADYHNHSNIETRLRKGGRTRCFGGCVFAYVGCYNKR
AYWVPRASADIGSGHTGITGDNVETLNEDLLEILSRERVNINIVGDFHLNEEVAIILASF
SASTSAFIDTIKSLDYKSFKTIVESCGNYKVTKGKPVKGAWNIGQQRSVLTPLCGFPSQA
AGVIRSIFARTLDAANHSIPDLQRAAVTILDGISEQSLRLVDAMVYTSDLLTNSVIIMAY
VTGGLVQQTSQWLSNLLGTTVEKLRPIFEWIEAKLSAGVEFLKDAWEILKFLITGVFDIV
KGQIQVASDNIKDCVKCFIDVVNKALEMCIDQVTIAGAKLRSLNLGEVFIAQSKGLYRQC
IRGKEQLQLLMPLKAPKEVTFLEGDSHDTVLTSEEVVLKNGELEALETPVDSFTNGAIVG
TPVCVNGLMLLEIKDKEQYCALSPGLLATNNVFRLKGGAPIKGVTFGEDTVWEVQGYKNV
RITFELDERVDKVLNEKCSVYTVESGTEVTEFACVVAEAVVKTLQPVSDLLTNMGIDLDE
WSVATFYLFDDAGEENFSSRMYCSFYPPDEEEEDDAECEEEEIDETCEHEYGTEDDYQGL
PLEFGASAETVRVEEEEEEDWLDDTTEQSEIEPEPEPTPEEPVNQFTGYLKLTDNVAIKC
VDIVKEAQSANPMVIVNAANIHLKHGGGVAGALNKATNGAMQKESDDYIKLNGPLTVGGS
CLLSGHNLAKKCLHVVGPNLNAGEDIQLLKAAYENFNSQDILLAPLLSAGIFGAKPLQSL

(continued)

Box 1.1 (continued)

QVCVQTVRTQVYIAVNDKALYEQVVMDYLDNLKPRVEAPKQEEPPNTEDSKTEEKSVVQK
PVDVKPKIKACIDEVTTTLEETKFLTNKLLLFADINGKLYHDSQNMLRGEDMSFLEKDAP
YMVGDVITSGDITCVVIPSKKAGGTTEMLSRALKKVPVDEYITTYPGQGCAGYTLEEAKT
ALKKCKSAFYVLPSEAPNAKEEILGTVSWNLREMLAHAEETRKLMPICMDVRAIMATIQR
KYKGIKIQEGIVDYGVRFFFYTSKEPVASIITKLNSLNEPLVTMPIGYVTHGFNLEEAAR
CMRSLKAPAVVSVSSPDAVTTYNGYLTSSSKTSEEHFVETVSLAGSYRDWSYSGQRTELG
VEFLKRGDKIVYHTLESPVEFHLDGEVLSLDKLKSLLSLREVKTIKVFTTVDNTNLHTQL
VDMSMTYGQQFGPTYLDGADVTKIKPHVNHEGKTFFVLPSDDTLRSEAFEYYHTLDESFL
GRYMSALNHTKKWKFPQVGGLTSIKWADNNCYLSSVLLALQQLEVKFNAPALQEAYYRAR
AGDAANFCALILAYSNKTVGELGDVRETMTHLLQHANLESAKRVLNVVCKHCGQKTTTLT
GVEAVMYMGTLSYDNLKTGVSIPCVCGRDATQYLVQQESSFVMMSAPPAEYKLQQGTFLC
ANEYTGNYQCGHYTHITAKETLYRIDGAHLTKMSEYKGPVTDVFYKETSYTTTIKPVSYK
LDGVTYTEIEPKLDGYYKKDNAYYTEQPIDLVPTQPLPNASFDNFKLTCSNTKFADDLNQ
MTGFTKPASRELSVTFFPDLNGDVVAIDYRHYSASFKKGAKLLHKPIVWHINQATTKTTF
KPNTWCLRCLWSTKPVDTSNSFEVLAVEDTQGMDNLACESQQPTSEEVVENPTIQKEVIE
CDVKTTEVVGNVILKPSDEGVKVTQELGHEDLMAAYVENTSITIKKPNELSLALGLKTIA
THGIAAINSVPWSKILAYVKPFLGQAAITTSNCAKRLAQRVFNNYMPYVFTLLFQLCTFT
KSTNSRIRASLPTTIAKNSVKSVAKLCLDAGINYVKSPKFSKLFTIAMWLLLLSICLGSL
ICVTAAFGVLLSNFGAPSYCNGVRELYLNSSNVTTMDFCEGSFPCSICLSGLDSLDSYPA
LETIQVTISSYKLDLTILGLAAEWVLAYMLFTKFFYLLGLSAIMQVFFGYFASHFISNSW
LMWFIISIVQMAPVSAMVRMYIFFASFYYIWKSYVHIMDGCTSSTCMMCYKRNRATRVEC
TTIVNGMKRSFYVYANGGRGFCKTHNWNCLNCDTFCTGSTFISDEVARDLSLQFKRPINP
TDQSSYIVDSVAVKNGALHLYFDKAGQKTYERHPLSHFVNLDNLRANNTKGSLPINVIVF
DGKSKCDESASKSASVYYSQLMCQPILLLDQALVSDVGDSTEVSVKMFDAYVDTFSATFS
VPMEKLKALVATAHSELAKGVALDGVLSTFVSAARQGVVDTDVDTKDVIECLKLSHHSDL
EVTGDSCNNFMLTYNKVENMTPRDLGACIDCNARHINAQVAKSHNVSLIWNVKDYMSLSE
QLRKQIRSAAKKNNIPFRLTCATTRQVVNVITTKISLKGGKIVSTCFKLMLKATLLCVLA
ALVCYIVMPVHTLSIHDGYTNEIIGYKAIQDGVTRDIISTDDCFANKHAGFDAWFSQRGG
SYKNDKSCPVVAAIITREIGFIVPGLPGTVLRAINGDFLHFLPRVFSAVGNICYTPSKLI
EYSDFATSACVLAAECTIFKDAMGKPVPYCYDTNLLEGSISYSELRPDTRYVLMDGSIIQ
FPNTYLEGSVRVVTTFDAEYCRHGTCERSEVGICLSTSGRWVLNNEHYRALSGVFCGVDA
MNLIANIFTPLVQPVGALDVSASVVAGGIIAILVTCAAYYFMKFRRVFGEYNHVVAANAL
LFLMSFTILCLVPAYSFLPGVYSVFYLYLTFYFTNDVSFLAHLQWFAMFSPIVPFWITAI
YVFCISLKHCHWFFNNYLRKRVMFNGVTFSTFEEAALCTFLLNKEMYLKLRSETLLPLTQ
YNRYLALYNKYKYFSGALDTTSYREAACCHLAKALNDFSNSGADVLYQPPQTSITSAVLQ
SGFRKMAFPSGKVEGCMVQVTCGTTTLNGLWLDDTVYCPRHVICTAEDMLNPNYEDLLIR
KSNHSFLVQAGNVQLRVIGHSMQNCLLRLKVDTSNPKTPKYKFVRIQPGQTFSVLACYNG
SPSGVYQCAMRPNHTIKGSFLNGSCGSVGFNIDYDCVSFCYMHHMELPTGVHAGTDLEGK
FYGPFVDRQTAQAAGTDTTITLNVLAWLYAAVINGDRWFLNRFTTTLNDFNLVAMKYNYE
PLTQDHVDILGPLSAQTGIAVLDMCAALKELLQNGMNGRTILGSTILEDEFTPFDVVRQC
SGVTFQGKFKKIVKGTHHWMLLTFLTSLLILVQSTQWSLFFFVYENAFLPFTLGIMAIAA
CAMLLVKHKHAFLCLFLLPSLATVAYFNMVYMPASWVMRIMTWLELADTSLSGYRLKDCV
MYASALVLLILMTARTVYDDAARRVWTLMNVITLVYKVYYGNALDQAISMWALVISVTSN
YSGVVTTIMFLARAIVFVCVEYYPLLFITGNTLQCIMLVYCFLGYCCCCYFGLFCLLNRY
FRLTLGVYDYLVSTQEFRYMNSQGLLPPKSSIDAFKLNIKLLGIGGKPCIKVATVQSKMS

Box 1.1 (continued)

DVKCTSVVLLSVLQQLRVESSSKLWAQCVQLHNDILLAKDTTEAFEKMVSLLSVLLSMQG
AVDINRLCEEMLDNRATLQAIASEFSSLPSYAAYATAQEAYEQAVANGDSEVVLKKLKKS
LNVAKSEFDRDAAMQRKLEKMADQAMTQMYKQARSEDKRAKVTSAMQTMLFTMLRKLDND
ALNNIINNARDGCVPLNIIPLTTAAKLMVVVPDYGTYKNTCDGNTFTYASALWEIQQVVD
ADSKIVQLSEINMDNSPNLAWPLIVTALRANSAVKLQNNELSPVALRQMSCAAGTTQTAC
TDDNALAYYNNSKGGRFVLALLSDHQDLKWARFPKSDGTGTIYTELEPPCRFVTDTPKGP
KVKYLYFIKGLNNLNRGMVLGSLAATVRLQAGNATEVPANSTVLSFCAFAVDPAKAYKDY
LASGGQPITNCVKMLCTHTGTGQAITVTPEANMDQESFGGASCCLYCRCHIDHPNPKGFC
DLKGKYVQIPTTCANDPVGFTLRNTVCTVCGMWKGYGCSCDQLREPLMQSADASTFLNRV
CGVSAARLTPCGTGTSTDVVYRAFDIYNEKVAGFAKFLKTNCCRFQEKDEEGNLLDSYFV
VKRHTMSNYQHEETIYNLVKDCPAVAVHDFFKFRVDGDMVPHISRQRLTKYTMADLVYAL
RHFDEGNCDTLKEILVTYNCCDDDYFNKKDWYDFVENPDILRVYANLGERVRQSLLKTVQ
FCDAMRDAGIVGVLTLDNQDLNGNWYDFGDFVQVAPGCGVPIVDSYYSLLMPILTLTRAL
AAESHMDADLAKPLIKWDLLKYDFTEERLCLFDRYFKYWDQTYHPNCINCLDDRCILHCA
NFNVLFSTVFPPTSFGPLVRKIFVDGVPFVVSTGYHFRELGVVHNQDVNLHSSRLSFKEL
LVYAADPAMHAASGNLLLDKRTTCFSVAALTNNVAFQTVKPGNFNKDFYDFAVSKGFFKE
GSSVELKHFFFAQDGNAAISDYDYYRYNLPTMCDIRQLLFVVEVVDKYFDCYDGGCINAN
QVIVNNLDKSAGFPFNKWGKARLYYDSMSYEDQDALFAYTKRNVIPTITQMNLKYAISAK
NRARTVAGVSICSTMTNRQFHQKLLKSIAATRGATVVIGTSKFYGGWHNMLKTVYSDVET
PHLMGWDYPKCDRAMPNMLRIMASLVLARKHNTCCNLSHRFYRLANECAQVLSEMVMCGG
SLYVKPGGTSSGDATTAYANSVFNICQAVTANVNALLSTDGNKIADKYVRNLQHRLYECL
YRNRDVDHEFVDEFYAYLRKHFSMMILSDDAVVCYNSNYAAQGLVASIKNFKAVLYYQNN
VFMSEAKCWTETDLTKGPHEFCSQHTMLVKQGDDYVYLPYPDPSRILGAGCFVDDIVKTD
GTLMIERFVSLAIDAYPLTKHPNQEYADVFHLYLQYIRKLHDELTGHMLDMYSVMLTNDN
TSRYWEPEFYEAMYTPHTVLQAVGACVLCNSQTSLRCGACIRRPFLCCKCCYDHVISTSH
KLVLSVNPYVCNAPGCDVTDVTQLYLGGMSYYCKSHKPPISFPLCANGQVFGLYKNTCVG
SDNVTDFNAIATCDWTNAGDYILANTCTERLKLFAAETLKATEETFKLSYGIATVREVLS
DRELHLSWEVGKPRPPLNRNYVFTGYRVTKNSKVQIGEYTFEKGDYGDAVVYRGTTTYKL
NVGDYFVLTSHTVMPLSAPTLVPQEHYVRITGLYPTLNISDEFSSNVANYQKVGMQKYST
LQGPPGTGKSHFAIGLALYYPSARIVYTACSHAAVDALCEKALKYLPIDKCSRIIPARAR
VECFDKFKVNSTLEQYVFCTVNALPETTADIVVFDEISMATNYDLSVVNARLRAKHYVYI
GDPAQLPAPRTLLTKGTLEPEYFNSVCRLMKTIGPDMFLGTCRRCPAEIVDTVSALVYDN
KLKAHKDKSAQCFKMFYKGVITHDVSSAINRPQIGVVREFLTRNPAWRKAVFISPYNSQN
AVASKILGLPTQTVDSSQGSEYDYVIFTQTTETAHSCNVNRFNVAITRAKIGILCIMSDR
DLYDKLQFTSLEIPRRNVATLQAENVTGLFKDCSKIITGLHPTQAPTHLSVDIKFKTEGL
CVDIPGIPKDMTYRRLISMMGFKMNYQVNGYPNMFITREEAIRHVRAWIGFDVEGCHATR
DAVGTNLPLQLGFSTGVNLVAVPTGYVDTENNTEFTRVNAKPPPGDQFKHLIPLMYKGLP
WNVVRIKIVQMLSDTLKGLSDRVVFVLWAHGFELTSMKYFVKIGPERTCCLCDKRATCFS
TSSDTYACWNHSVGFDYVYNPFMIDVQQWGFTGNLQSNHDQHCQVHGNAHVASCDAIMTR
CLAVHECFVKRVDWSVEYPIIGDELRVNSACRKVQHMVVKSALLADKFPVLHDIGNPKAI
KCVPQAEVEWKFYDAQPCSDKAYKIEELFYSYATHHDKFTDGVCLFWNCNVDRYPANAIV
CRFDTRVLSNLNLPGCDGGSLYVNKHAFHTPAFDKSAFTNLKQLPFFYYSDSPCESHGKQ
VVSDIDYVPLKSATCITRCNLGGAVCRHHANEYRQYLDAYNMMISAGFSLWIYKQFDTYN
LWNTFTRLQSLENVAYNVVNKGHFDGHAGEAPVSIINNAVYTKVDGIDVEIFENKTTLPV
NVAFELWAKRNIKPVPEIKILNNLGVDIAANTVIWDYKREAPAHVSTIGVCTMTDIAKKP

(continued)

Box 1.1 (continued)

TESACSSLTVLFDGRVEGQVDLFRNARNGVLITEGSVKGLTPSKGPAQASVNGVTLIGES
VKTQFNYFKKVDGIIQQLPETYFTQSRDLEDFKPRSQMETDFLELAMDEFIQRYKLEGYA
FEHIVYGDFSHGQLGGLHLMIGLAKRSQDSPLKLEDFIPMDSTVKNYFITDAQTGSSKCV
CSVIDLLLDDFVEIIKSQDLSVISKVVKVTIDYAEISFMLWCKDGHVETFYPKLQASQAW
QPGVAMPNLYKMQRMLLEKCDLQNYGENAVIPKGIMMNVAKYTQLCQYLNTLTLAVPYNM
RVIHFGAGSDKGVAPGTAVLRQWLPTGTLLVDSDLNDFVSDADSTLIGDCATVHTANKWD
LIISDMYDPRTKHVTKENDSKEGFFTYLCGFIKQKLALGGSIAVKITEHSWNADLYKLMG
HFSWWTAFVTNVNASSSEAFLIGANYLGKPKEQIDGYTMHANYIFWRNTNPIQLSSYSLF
DMSKFPLKLRGTAVMSLKENQINDMIYSLLEKGRLIIRENNRVVVSSDILVNN

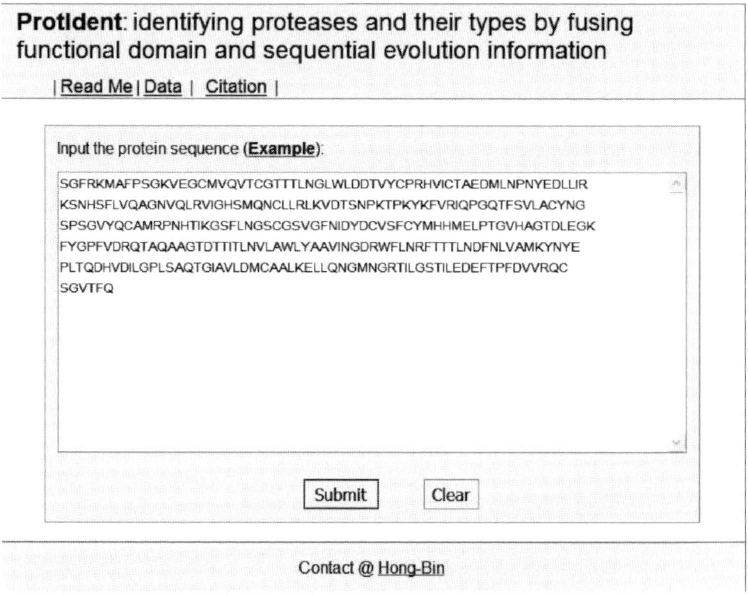

Fig. 1.10 A screenshot to show the sequence of the SARS main protease entered into the input box of the ProtIdent web server

Step 3. Click on the Submit button and you will see in about 15 s that "the query protein is a protease" and it "belongs to Cysteine type", as shown on the output screen of Fig. 1.11. If your input is the 7,073 amino acids of the full length SARS coronavirus protease as taken from **Box 1.1**, the same predicted result will be obtained although it will take much longer time (about 230 s) because the input sequence now is much longer.

Step 4. Click on the Citation button to find the relevant papers that document the detailed development and algorithm of ProtIdent.
Step 5. Click on the Data button to download the benchmark datasets used to train and test the ProtIdent predictor.

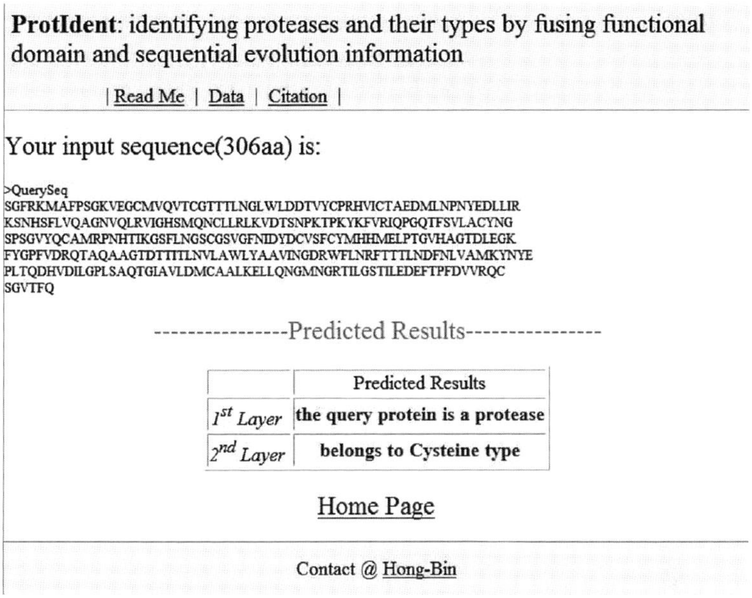
ProtIdent: identifying proteases and their types by fusing functional domain and sequential evolution information

| Read Me | Data | Citation |

Your input sequence(306aa) is:

>QuerySeq
SGFRKMAFPSGKVEGCMVQVTCGTTTLNGLWLDDTVYCPRHVICTAEDMLNPNYEDLLIR
KSNHSFLVQAGNVQLRVIGHSMQNCLLRLKVDTSNPKTPKYKFVRIQPGQTFSVLACYNG
SPSGVYQCAMRPNHTIKGSFLNGSCGSVGFNIDYDCVSFCYMHHMELPTGVHAGTDLEGK
FYGPFVDRQTAQAAGTDTTITLNVLAWLYAAVINGDRWFLNRFTTTLNDFNLVAMKYNYE
PLTQDHVDILGPLSAQTGIAVLDMCAALKELLQNGMNGRTILGSTILEDEFTPFDVVRQC
SGVTFQ

----------------Predicted Results----------------

	Predicted Results
1st Layer	**the query protein is a protease**
2nd Layer	**belongs to Cysteine type**

Home Page

Contact @ Hong-Bin

Fig. 1.11 A screenshot to show the output predicted by the ProtIdent web server

Caveat. To obtain the predicted result with the expected success rate, either the entire sequence of the query protein or the entire sequence of its main protease should be used as an input.

1.8 Conclusion

Since the outbreak of SARS in November 2002 in Southern China's Guangdong Province, considerable progress has been made in developing drugs for SARS therapy. This chapter is focused on the advances achieved mainly from the structural bioinformatics, pharmacophore modeling, molecular docking, peptide-cleavage site prediction, and other computational approaches. It is indicated that the compounds $C_{28}H_{34}O_4N_7Cl$, $C_{21}H_{36}O_5N_6$ and $C_{21}H_{36}O_5N_6$ (Wei et al., 2006), as well as KZ7088 (Chou et al., 2003), a derivative of AG7088, might be the promising candidates for further investigation. Meanwhile, it is elucidated according to the "distorted key" theory (Chou et al., 1996) that the octapeptides ATLQAENV and AVLQSGFR may be converted to effective inhibitors against the SARS enzyme. Also, how to modify these octapeptides and make them become potent inhibitors is suggested. Finally, a step-to-step guide on how to use ProtIdent, a web server for identifying proteases and their types (Chou and Shen, 2008), is presented that may be useful for people working in the area of Proteases in Biology and Disease.

References

Adessi, C. and Soto, C. 2002, Converting a peptide into a drug: strategies to improve stability and bioavailability. *Curr Med Chem*, **9**, 963–978.

Anand, K., Ziebuhr, J., Wadhwani, P., Mesters, J.R. and Hilgenfeld, R. 2003, Coronavirus main proteinase (3CLpro) structure: basis for design of anti-SARS drugs. *Science*, **300**, 1763–1767.

Baurin, N. 2002, Etude et développement de techniques QSAR pour la recherche de molécules d'intérêts thérapeutique. *Criblage virtuel et Analyse de chmiothèque, Université d'Orléans*.

Benkirane, N., Guichard, G., Briand, J.P. and Muller, S. 1996, Exploration of requirements for peptidomimetic immune recognition. Antigenic and immunogenic properties of reduced peptide bond pseudopeptide analogues of a histone hexapeptide. *J Biol Chem*, **271**, 33218–33224.

Cai, Y.D. and Chou, K.C. 1998, Artificial neural network model for HIV protease cleavage sites in proteins. *Adv Eng Softw*, **29**, 119–128.

Chen, L.L., Ou, H.Y., Zhang, R. and Zhang, C.T. 2003, ZCURVE_CoV: a new system to recognize protein coding genes in coronavirus genomes, and its applications in analyzing SARS-CoV genomes. *Biochem Biophys Res Commun*, **307**, 382–388.

Chou, J.J. 1993a, Predicting cleavability of peptide sequences by HIV protease via correlation-angle approach. *J Protein Chem*, **12**, 291–302.

Chou, K.C. 1993b, A vectorized sequence-coupling model for predicting HIV protease cleavage sites in proteins. *J Biol Chem*, **268**, 16938–16948.

Chou, K.C. 1996, Review: prediction of HIV protease cleavage sites in proteins. *Analyt Biochem*, **233**, 1–14.

Chou, K.C. 2004a, Molecular therapeutic target for type-2 diabetes. *J Proteome Res*, **3**, 1284–1288.

Chou, K.C. 2004b, Review: structural bioinformatics and its impact to biomedical science. *Curr Med Chem*, **11**, 2105–2134.

Chou, K.C. 2004c, Insights from modelling the 3D structure of the extracellular domain of alpha7 nicotinic acetylcholine receptor. *Biochem Biophys Res Commun*, **319**, 433–438.

Chou, K.C. 2005a, Coupling interaction between thromboxane A2 receptor and alpha-13 subunit of guanine nucleotide-binding protein. *J Proteome Res*, **4**, 1681–1686.

Chou, K.C. 2005b, Modeling the tertiary structure of human cathepsin-E. *Biochem Biophys Res Commun*, **331**, 56–60.

Chou, K.C. and Howe, W.J. 2002, Prediction of the tertiary structure of the beta-secretase zymogen. *Biochem Biophys Res Commun*, **292**, 702–708.

Chou, K.C. and Shen, H.B. 2008, ProtIdent: a web server for identifying proteases and their types by fusing functional domain and sequential evolution information. *Biochem Biophys Res Commun*, **376**, 321–325.

Chou, K.C., Zhang, C.T. and Kezdy, F.J. 1993, A vector approach to predicting HIV protease cleavage sites in proteins. *Proteins: Struct Funct Genet*, **16**, 195–204.

Chou, K.C., Tomasselli, A.L., Reardon, I.M. and Heinrikson, R.L. 1996, Predicting HIV protease cleavage sites in proteins by a discriminant function method. *Proteins: Struct Funct Genet*, **24**, 51–72.

Chou, K.C., Jones, D. and Heinrikson, R.L. 1997, Prediction of the tertiary structure and substrate binding site of caspase-8. *FEBS Lett*, **419**, 49–54.

Chou, K.C., Watenpaugh, K.D. and Heinrikson, R.L. 1999, A Model of the complex between cyclin-dependent kinase 5(Cdk5) and the activation domain of neuronal Cdk5 activator. *Biochem Biophys Res Commun*, **259**, 420–428.

Chou, K.C., Tomasselli, A.G. and Heinrikson, R.L. 2000, Prediction of the tertiary structure of a caspase-9/inhibitor complex. *FEBS Lett*, **470**, 249–256.

Chou, K.C., Wei, D.Q. and Zhong, W.Z. 2003, Binding mechanism of coronavirus main proteinase with ligands and its implication to drug design against SARS. (Erratum: ibid., 2003, 310, 675). *Biochem Biophys Res Commun*, **308**, 148–151.

Clercq, E.D. 2006, Potential antivirals and antiviral strategies against SARS coronavirus infections. *Expert Rev Anti-infect Ther*, **4**, 291–302.

Connolly, M.L. 1993, The molecular surface package. *J Mol Graph*, **11**, 139–141.

Du, Q.S., Wang, S.Q., Wei, D.Q., Zhu, Y., Guo, H., Sirois, S. and Chou, K.C. 2004, Polyprotein Cleavage Mechanism of SARS CoV Mpro and Chemical Modification of Octapeptide. *Peptides*, **25**, 1857–1864.

Du, Q.S., Wang, S., Wei, D.Q., Sirois, S. and Chou, K.C. 2005a, Molecular modelling and chemical modification for finding peptide inhibitor against SARS CoV Mpro. *Analyt Biochem*, **337**, 262–270.

Du, Q.S., Wang, S.Q., Jiang, Z.Q., Gao, W.N., Li, Y.D., Wei, D.Q. and Chou, K.C. 2005b, Application of bioinformatics in search for cleavable peptides of SARS-CoV Mpro and chemical modification of octapeptides. *Med Chem*, **1**, 209–213.

Du, Q.S., Wang, S.Q. and Chou, K.C. 2007a, Analogue inhibitors by modifying oseltamivir based on the crystal neuraminidase structure for treating drug-resistant H5N1 virus. *Biochem Biophys Res Commun*, **362**, 525–531.

Du, Q.S., Sun, H. and Chou, K.C. 2007b, Inhibitor design for SARS coronavirus main protease based on "distorted key theory". *Med Chem*, **3**, 1–6.

Gan, Y.R., Huang, H., Huang, Y.D., Rao, C.M., Zhao, Y., Liu, J.S., Wu, L. and Wei, D.Q. 2006, Synthesis and activity assess of an octapeptide inhibitor designed for SARS coronavirus main proteinase. *Peptides*, **27**, 622–625.

Gao, F., Ou, H.Y., Chen, L.L., Zheng, W.X. and Zhang, C.T. 2003, Prediction for proteinase cleavage sites in polyproteins of coronaviruses and its applications in analyzing SARS-CoV genomes. *FEBS Letters*, **553**, 451–456.

Gao, L., Ding, Y.S., Dai, H., Shao, S.H., Huang, Z.D. and Chou, K.C. 2006, A novel fingerprint map for detecting SARS-CoV. *J Pharmaceut Biomed Analysis*, **41**, 246–250.

Gao, W.N., Wei, D.Q., Li, Y., Gao, H., Xu, W.R., Li, A.X. and Chou, K.C. 2007, Agaritine and its derivatives are potential inhibitors against HIV proteases. *Med Chem*, **3**, 221–226.

Gong, K., Li, L., Wang, J.F., Cheng, F., Wei, D.Q. and Chou, K.C. 2009, Binding mechanism of H5N1 influenza virus neuraminidase with ligands and its implication for drug design. *Med Chem*, in press.

Graham, S.L., deSolms, S.J., Giuliani, E.A., Kohl, N.E., Mosser, S.D., Oliff, A.I., Pompliano, D.L., Rands, E., Breslin, M.J., Deana, A.A. et al. 1994, Pseudopeptide inhibitors of Ras farnesyl-protein transferase. *J Med Chem*, **37**, 725–732.

Gu, R.X., Gu, H., Xie, Z.Y., Wang, J.F., Arias, H.R., Wei, D.Q. and Chou, K.C. 2009, Possible drug candidates for Alzheimer's disease deduced from studying their binding interactions with alpha7 nicotinic acetylcholine receptor. *Med Chem*, in press.

Guo, X.L., Li, L., Wei, D.Q., Zhu, Y.S. and Chou, K.C. 2008, Cleavage mechanism of the H5N1 hemagglutinin by trypsin and furin. *Amino Acids*, **35**, 375–382.

Kesel, A.J. 2005, Synthesis of novel test compounds for antiviral chemotherapy of severe acute respiratory syndrome (SARS). *Curr Med Chem*, **12**, 2095–2162.

Li, L., Wei, D.Q., Wang, J.F. and Chou, K.C. 2007a, Computational studies of the binding mechanism of calmodulin with chrysin. *Biochem Biophys Res Comm*, **358**, 1102–1107.

Li, Y., Wei, D.Q., Gao, W.N., Gao, H., Liu, B.N., Huang, C.J., Xu, W.R., Liu, D.K., Chen, H.F. and Chou, K.C. 2007b, Computational approach to drug design for oxazolidinones as antibacterial agents. *Med Chem*, **3**, 576–582.

Liang, G.Z. and Li, S.Z. 2007, A new sequence representation as applied in better specificity elucidation for human immunodeficiency virus type 1 protease. *Biopolymers*, **88**, 401–412.

Lipinski, C.A. 2000, Drug-like properties and the causes of poor solubility and poor permeability. *J Pharmacol Toxicol Methods*, **44**, 235–249.

Lowther, W.T., Majer, P. and Dunn, B.M. 1995, Engineering the substrate specificity of rhizopuspepsin: the role of Asp 77 of fungal aspartic proteinases in facilitating the cleavage of oligopeptide substrates with lysine in P1. *Protein Sci*, **4**, 689–702.

Miller, M., Schneider, J., Sathyanarayana, B.K., Toth, M.V., Marshall, G.R., Clawson, L., Selk, L., Kent, S.B. and Wlodawer, A. 1989, Structure of complex of synthetic HIV-1 protease with a substrate-based inhibitor at 2.3 A resolution. *Science*, **246**, 1149–1152.

Oprea, T.I. 2000, Property distribution of drug-related chemical databases. *J Comput Aided Mol Des*, **14**, 251–264.
Peiris, J.S., Chu, C.M., Cheng, V.C., Chan, K.S., Hung, I.F., Poon, L.L., Law, K.I., Tang, B.S., Hon, T.Y., Chan, C.S. et al. 2003, Clinical progression and viral load in a community outbreak of coronavirus-associated SARS pneumonia: a prospective study. *Lancet*, **361**, 1767–1772.
Poorman, R.A., Tomasselli, A.G., Heinrikson, R.L. and Kezdy, F.J. 1991, A cumulative specificity model for proteases from human immunodeficiency virus types 1 and 2, inferred from statistical analysis of an extended substrate data base. *J Biol Chem*, **266**, 14554–14561.
Rishton, G.M. 1997, Reactive compounds and in vitro false positives in HTS. *Drug Discovery Today*, **2**, 382–384.
Rognvaldsson, T., You, L. and Garwicz, D. 2007, Bioinformatic approaches for modeling the substrate specificity of HIV-1 protease: an overview. *Expert Rev Mol Diagn*, **7**, 435–451.
Samee, W. 2005, Severe acute respiratory syndrome coronavirus (SARS-CoV) protease inhibitors. *J Club Pharm Chem Pharmacognosy*, **January**, 1–12.
Schechter, I. and Berger, A. 1967, On the size of the active site in protease. I. Papain. *Biochem Biophys Res Comm*, **27**, 157–162.
Schnell, J.R. and Chou, J.J. 2008, Structure and mechanism of the M2 proton channel of influenza A virus. *Nature*, **451**, 591–595.
Shen, H.B. and Chou, K.C. 2008, HIVcleave: a web-server for predicting HIV protease cleavage sites in proteins. *Analyt Biochem*, **375**, 388–390.
Sirois, S., Wei, D.Q., Du, Q.S. and Chou, K.C. 2004, Virtual Screening for SARS-CoV protease based on KZ7088 pharmacophore points. *J Chem Inf Comput Sci*, **44**, 1111–1122.
Szelke, M., Leckie, B.J., Tree, M., Brown, A., Grant, J., Hallett, A., Hughes, M., Jones, D.M. and Lever, A.F. 1982, A potent new renin inhibitor. In vitro and in vivo studies. *Hypertension*, **4**, 59.
Thompson, T.B., Chou, K.C. and Zheng, C. 1995, Neural network prediction of the HIV-1 protease cleavage sites. *J Theoret Biol*, **177**, 369–379.
Venkatesan, N., Kim, B.H. 2002, Synthesis and enzyme inhibitory activities of novel peptide isosteres. *Curr Med Chem*, **9**, 2243–2270.
Wang, J.F., Wei, D.Q., Li, L., Zheng, S.Y., Li, Y.X. and Chou, K.C. 2007a, 3D structure modeling of cytochrome P450 2C19 and its implication for personalized drug design. *Biochem Biophys Res Commun (Corrigendum: ibid, 2007, Vol357, 330)*, **355**, 513–519.
Wang, J.F., Wei, D.Q., Lin, Y., Wang, Y.H., Du, H.L., Li, Y.X. and Chou, K.C. 2007b, Insights from modeling the 3D structure of NAD(P)H-dependent D-xylose reductase of Pichia stipitis and its binding interactions with NAD and NADP. *Biochem Biophys Res Comm*, **359**, 323–329.
Wang, S.Q., Du, Q.S., Zhao, K., Li, A.X., Wei, D.Q. and Chou, K.C. 2007c, Virtual screening for finding natural inhibitor against cathepsin-L for SARS therapy. *Amino Acids*, **33**, 129–135.
Wang, S.Q., Du, Q.S. and Chou, K.C. 2007d, Study of drug resistance of chicken influenza A virus (H5N1) from homology-modeled 3D structures of neuraminidases. *Biochem Biophys Res Comm*, **354**, 634–640.
Wang, J.F., Wei, D.Q., Chen, C., Li, Y. and Chou, K.C. 2008, Molecular modeling of two CYP2C19 SNPs and its implications for personalized drug design. *Protein Pept Lett*, **15**, 27–32.
Wang, M., Yao, J.S., Huang, Z.D., Xu, Z.J., Liu, G.P., Zhao, H.Y., Wang, X.Y., Yang, J., Zhu, Y.S. and Chou, K.C. 2005, A new nucleotide-composition based fingerprint of SARS-CoV with visualization analysis. *Med Chem*, **1**, 39–47.
Wei, D.Q., Sirois, S., Du, Q.S., Arias, H.R. and Chou, K.C. 2005, Theoretical studies of Alzheimer's disease drug candidate [(2,4-dimethoxy) benzylidene]-anabaseine dihydrochloride (GTS-21) and its derivatives. *BBRC*, **338**, 1059–1064.
Wei, D.Q., Du, Q.S., Sun, H. and Chou, K.C. 2006a, Insights from modeling the 3D structure of H5N1 influenza virus neuraminidase and its binding interactions with ligands. *Biochem Biophys Res Comm*, **344**, 1048–1055.
Wei, D.Q., Zhang, R., Du, Q.S., Gao, W.N., Li, Y., Gao, H., Wang, S.Q., Zhang, X., Li, A.X., Sirois, S. et al. 2006b, Anti-SARS drug screening by molecular docking. *Amino Acids*, **31**, 73–80.

Wei, H., Zhang, R., Wang, C., Zheng, H., Chou, K.C. and Wei, D.Q. 2007, Molecular insights of SAH enzyme catalysis and their implication for inhibitor design. *J Theoret Biol*, **244**, 692–702.

Wolfram, S. 1984, Cellular automation as models of complexity. *Nature*, **311**, 419–424.

Wolfram, S. 2002, *A New Kind of Science*. Wolfram Media, Champaign, IL.

Wu, Y.S., Lin, W.H., Hsu, J.T. and Hsieh, H.P. 2006, Antiviral drug discovery against SARS-CoV. *Curr Med Chem*, **13**, 2003–2020.

Xu, J. and Stevenson, J. 2000, Drug-like index: a new approach to measure drug-like compounds and their diversity. *J Chem Inf Comput Sci*, **40**, 1177–1187.

Yang, H., Yang, M., Ding, Y., Liu, Y., Lou, Z., Zhou, Z., Sun, L., Mo, L., Ye, S., Pang, H. et al. 2003, The crystal structures of severe acute respiratory syndrome virus main protease and its complex with an inhibitor. *Proc Natl Acad Sci USA*, **100**, 13190–13195.

You, L., Garwicz, D. and Rognvaldsson, T. 2005, Comprehensive bioinformatic analysis of the specificity of human immunodeficiency virus type 1 protease. *J Virol*, **79**, 12477–12486.

Zhang, C.T. and Chou, K.C. 1993, An alternate-subsite-coupled model for predicting HIV protease cleavage sites in proteins. *Protein Eng*, **7**, 65–73.

Zhang, R., Wei, D.Q., Du, Q.S. and Chou, K.C. 2006, Molecular modeling studies of peptide drug candidates against SARS. *Med Chem*, **2**, 309–314.

Zhang, X.W. and Yap, Y.L. 2004, Exploring the binding mechanism of the main proteinase in SARS-associated coronavirus and its implication to anti-SARS drug design. *Bioorg Med Chem*, **12**, 2219–2223.

Zheng, H., Wei, D.Q., Zhang, R., Wang, C., Wei, H. and Chou, K.C. 2007, Screening for new agonists against Alzheimer's disease. *Med Chem*, **3**, 488–493.

Zhou, G.P. and Troy, F.A., 2nd ed. 2003, Characterization by NMR and molecular modeling of the binding of polyisoprenols and polyisoprenyl recognition sequence peptides: 3D structure of the complexes reveals sites of specific interactions. *Glycobiology*, **13**, 51–71.

Zhou, G.P. and Troy, F.A., 2nd ed. 2005a, NMR study of the preferred membrane orientation of polyisoprenols (dolichol) and the impact of their complex with polyisoprenyl recognition sequence peptides on membrane structure. *Glycobiology*, **15**, 347–359.

Zhou, G.P. and Troy, F.A. 2005b, NMR studies on how the binding complex of polyisoprenol recognition sequence peptides and polyisoprenols can modulate membrane structure. *Curr Protein Peptide Sci*, **6**, 399–411.

Chapter 2
HIV-1 Protease and AIDS Therapy

Irene T. Weber, Ying Zhang, and Jozsef Tözsér

Abstract Infection with HIV-1 causes the pandemic disease of AIDS in an estimated 30 million people. The viral protease has proved a successful target for AIDS therapy, and current effective therapy uses a selection from eight protease inhibitors in combination with other antiviral drugs. The antiviral protease inhibitors were developed over the past 15 years based on knowledge of the molecular structure of the protease, its substrate specificity and the mode of binding of inhibitors. Today, drug resistance has become a major challenge in AIDS therapy. Consequently, in the combat against drug resistance new antiviral protease inhibitors are being developed from structure-guided designs. These new drugs, such as darunavir, demonstrate high affinity binding to resistant mutants of HIV protease, and provide high genetic barriers to resistance. Recent developments are described in these structure-guided designs for antiviral therapy based on targeting the HIV-1 protease and its drug resistant mutants.

Keywords HIV-1 protease • AIDS therapy • protease inhibitors • drug resistance

I.T. Weber (✉)
Department of Biology, Molecular Basis of Disease Program, Georgia State University, 24 Peachtree Center Avenue, Atlanta, GA 30303, USA; Department of Chemistry, Molecular Basis of Disease Program, Georgia State University, Atlanta, GA 30303, USA
e-mail: bioitw@langate.gsu.edu

Y. Zhang
Department of Chemistry, Molecular Basis of Disease Program, Georgia State University, Atlanta, GA 30303, USA

J. Tözsér
Department of Biology, Molecular Basis of Disease Program, Georgia State University, 24 Peachtree Center Avenue, Atlanta, GA 30303, USA; Department of Biochemistry and Molecular Biology, Research Center for Molecular Medicine, Medical and Health Science Center, University of Debrecen, Debrecen, Egyetem tér 1, Life Science Building, Hungary

U. Lendeckel and N.M. Hooper (eds.), *Viral Proteases and Antiviral Protease Inhibitor Therapy*, Proteases in Biology and Disease 8,

2.1 Introduction

The aspartic protease encoded by human immunodeficiency virus type 1 (HIV-1) is an important target for antiviral therapy for AIDS. The pandemic disease AIDS is caused by infection with the retrovirus HIV-1 and is usually lethal without treatment. Worldwide, more than 30 million people are suffering from HIV-1 infection. Treatment with a cocktail of antiviral drugs has proved effective in reducing viral load and the severity of disease, which has become a chronic condition with current therapy. Today, there are several types of anti-HIV drugs that target the three enzymes, reverse transcriptase, protease and integrase, as well as viral entry and fusion with cells. The recommended treatment regimen for HIV-1 infected patients is an individualized combination of drugs, usually including a protease inhibitor and two reverse transcriptase inhibitors (Hammer et al., 2008; Martinez-Cajas and Wainberg, 2008). The inclusion of protease inhibitors in 1995 heralded a new era in AIDS therapy with a dramatic increase in survival of AIDS patients. The clinical protease inhibitors were designed based on accumulated knowledge of the protease structure, substrate specificity and mode of binding, as well as understanding of the role of the protease in viral replication. In fact, the development of clinical inhibitors of HIV-1 protease has become a paradigm for structure-guided drug design (Wlodawer and Vondrasek, 1998). Currently, AIDS therapy faces the serious challenge of drug resistance due to the high genetic diversity and error-prone replication of the virus, and the therapeutic complications of poor patient compliance and toxic side effects. HIV-1 rapidly evolves resistance to single drugs. Hence, effective treatment of resistant HIV requires therapy with multiple drugs (Lagnese and Daar, 2008). New protease inhibitors have been developed to combat resistance. New drugs bind tightly to resistant mutants of HIV-1 protease, as well as to the wild type enzyme, and show higher genetic barriers to resistance (Ghosh et al., 2008a; Martinez-Cajas and Wainberg, 2007). Interestingly, HIV-1 protease inhibitors have shown promise for treatment of other diseases including bacterial and fungal infections as well as cancer (Mastrolorenzo et al., 2007). Here, we describe the basic properties of HIV-1 protease, its importance in structure-guided drug design for AIDS, and discuss recent developments in antiviral therapy based on targeting the HIV-1 protease and the drug resistant mutants.

2.2 Role of HIV-1 Protease in Viral Replication

HIV-1 is the best characterized human retrovirus due to intensive study over the past 25 years. HIV-1 is genetically heterogeneous comprising three major groups M, O and N, with group M infections predominating worldwide (Buonaguro et al., 2007). Group M is subdivided into ten subtypes with about 15–25% genetic variation. The genome of HIV-1 consists of genes for three polyprotein precursors Gag, Gag-Pro-Pol and Env, as well as several shorter accessory proteins (Tözsér and

Oroszlan, 2003). HIV-1 protease is encoded by the *pro* gene located between *gag* and *pol*, and the protease-containing polyprotein is synthesized by a ribosomal frameshift (Wilson et al., 1988). As a consequence of the low, 5–10%, frequency of frameshifting, the replication enzymes including the protease are produced in substantially lower amounts compared to the structural proteins encoded by the *gag* gene. In the normal viral life-cycle, the viral precursor proteins are synthesized, and then assembled at the cell membrane for budding and formation of the immature viral particles. After budding, HIV-1 protease is responsible for hydrolyzing the Gag and Gag-Pro-Pol precursors during viral maturation to produce the individual enzymes, protease, reverse transcriptase and integrase, as well as the structural proteins matrix, capsid and nucleocapsid (Fig. 2.1). The protease recognizes and hydrolyzes specific sites with specific amino acid sequences. The sequences of the protease cleavage sites in the viral precursor Gag and Gag-Pro-Pol polproteins are indicated in Fig. 2.1. The precursors are cleaved in a sequentially ordered fashion in studies using *in vitro* assay systems (Pettit et al., 1994, 2005). The importance of sequential cleavage has also been demonstrated in virions with mutated cleavage sites (Wiegers et al., 1998). The mature, fully-active protease dimer is released by autoproteolytic processing of a dimeric Gag-Pro-Pol precursor (Louis et al., 1999; Petit et al., 1994; 2005). It is difficult to estimate the exact order of cleavage in the virion, but studies indicate that the first site to be cleaved is p2/NC and one of the slowest sites is CA/p2 in Gag processing. Accurate and ordered processing of Gag and Gag-Pro-Pol precursors is required for production of infectious virus.

The assembly of virus particles is highly coordinated with polyprotein cleavage by the protease. Interfering with this process by inhibiting the protease, or the presence of protease mutants with sufficiently altered activity, will also alter viral

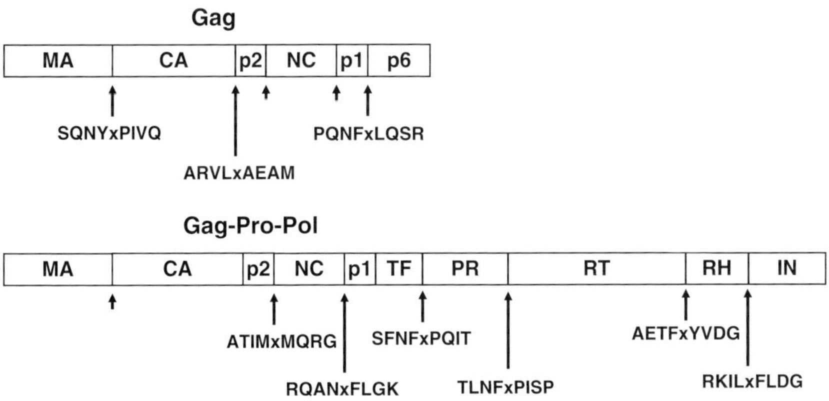

Fig. 2.1 HIV protease (PR) hydrolyzes the viral Gag and Gag-Pro-Pol precursor polyproteins during viral maturation to produce the mature viral enzymes, PR, reverse transcriptase (RT), integrase (IN), and the structural proteins matrix (MA), capsid (CA) and nucleocapsid (NC). The arrows indicate the amino acid sequences of the cleavage sites with x indicating the hydrolyzed peptide bond between P1 and P1′. Arrows without sequences indicate the same sequence as in the other precursor

replication so that no infectious virus can be produced. Consequently, inhibitors of HIV-1 protease are effective antiviral agents for AIDS therapy. In addition, the polyprotein substrate of the protease has become a new target for antiviral agents. A novel antiviral inhibitor of maturation is in clinical trials. Bevirimat targets the CA/p2 cleavage site and blocks Gag processing (Temesgen and Feinberg, 2006; Li et al., 2003). Other maturation inhibitors can be designed on the same principle to block maturation by binding to other critical cleavage sites. The majority of clinical antiviral inhibitors, however, bind in the protease active site and were designed with knowledge of the HIV-1 protease structure and substrate specificity.

2.3 Structure and Specificity of HIV-1 Protease

2.3.1 Critical Structural Features

HIV-1 protease is a member of the aspartic protease family of enzymes that share similar tertiary structures and amino acid sequences (Wlodawer and Gustchina, 2000). They share the same catalytic triplet of -Asp-Thr/Ser-Gly- and the characteristic properties of enzyme activity at acidic pH and inhibition by pepstatin. The subcategory of retroviral proteases share an additional characteristic triplet sequence of -Gly-Arg-Asp/Asn- and fold into similar tertiary and quaternary structures. HIV-1 protease forms a catalytically active dimer with 99 amino acid residues in each subunit. The protease dimer and critical structural regions: the active site, the dimer interface and flexible flaps, are indicated in Fig. 2.2. The two subunits, conventionally labeled as residues 1–99 and 1′–99′, have identical folds and the secondary structure comprises mostly beta strands and one short alpha-helix. The active-site triplet (-Asp25-Thr26-Gly27-) is highly conserved among aspartic proteases. Asp25 is essential for catalysis; mutation to Asn will inactivate the enzyme and destabilize the dimer (Sayer et al., 2008). The two catalytic Asp25 residues are arranged close together in the enzymatically active protease dimer. The two carboxylate side chains lie adjacent, in almost the same plane, and the closest oxygen atoms of the two residues are separated by 3.0 Å or less, which suggests protonation of one Asp25. The active site cavity is partially enclosed by the pair of flexible flaps, comprising the antiparallel beta-strands of residues 45–55. These flaps generally exist in more open conformations in the absence of bound substrate or inhibitor. The flaps contribute to binding of substrate or inhibitor and form intersubunit contacts between their tips around Ile50/50′. The short alpha-helix commences with the common retroviral protease sequence of -Gly-Arg-Asp/Asn-. Each subunit is stabilized internally by the aliphatic residues in adjacent hydrophobic regions (Leu23, Ile84, Val82, Val32, Leu76, Ile 47, Ile54, Val56) and (Val11, Ile13, Leu24, Leu90, Ile85, Ile64, Ile66, Leu89, Ile93, Ile15, Val75, Ile62, Leu38, Val77). The two subunits in the protease dimer interact in several regions: the tips of the flaps, the active site triplets, the ionic interactions among Asp29, Arg87 and Arg8′, and a major contribution from the four-stranded antiparallel beta sheet formed by the two N-termini and two C-termini (residues 1–4 and 96–99) as described (Weber, 1990).

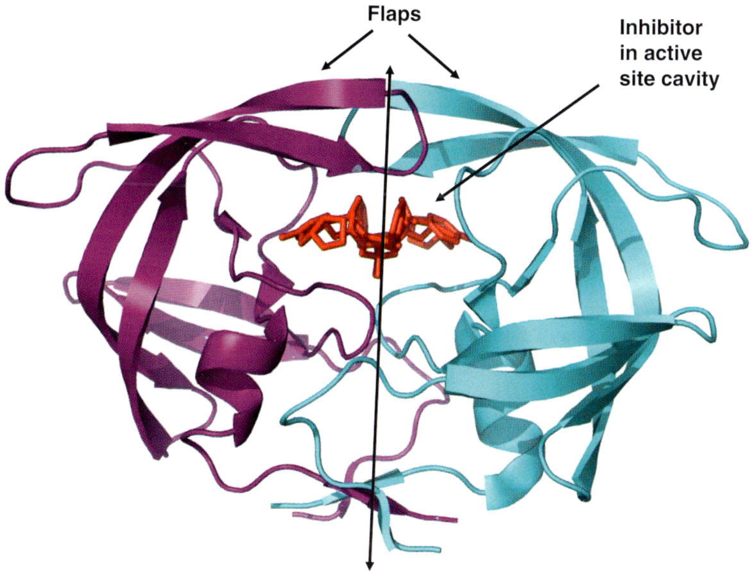

Fig. 2.2 Structure of HIV protease dimer. The protease backbone is shown in a ribbon representation colored magenta and cyan for the two subunits, with the inhibitor darunavir (red sticks) bound in the activity site cavity. The dimer interface and flexible flaps are labeled

2.3.2 *Recognition of Substrates*

HIV-1 protease recognizes and hydrolyzes the specific cleavage sites in the viral precursors or peptides with the same sequences (Fig. 2.1). The protease shows the most efficient catalysis on peptides spanning P4-P3′, where the peptide bond is cleaved between P1 and P1′ in the standard nomenclature for proteases. A number of early studies revealed the specificity of HIV-1 protease for peptides with various P4-P3′ sequences, as reviewed in (Louis et al., 2000; Beck et al., 2002). The specificity reflects the commonly observed amino acids at each position in the polyprotein cleavage sites. Retroviral protease cleavage sites are currently classified into two groups, although many cleavage sites do not belong in these groups. Type 1 cleavage sites have an aromatic residue at P1 and Pro at P1′, while type 2 sites have hydrophobic residues (excluding Pro) at the site of cleavage (P1 and P1′). The P2 and P2′ positions are also important in determining the specificity. In type 1 cleavage sites of primate lentiviruses, like HIV-1, there is a preference for Asn at P2 and beta-branched hydrophobic residues (Val or Ile) at P2′, while in type 2 cleavage sites the P2 position is typically beta-branched and the P2′ residue is Glu or Gln. Residues lying outside of the central P2–P2′ region can also substantially influence the rate of cleavage. This knowledge of substrate specificity has informed the design of antiviral inhibitors.

The molecular basis for protease specificity for viral polyprotein cleavage sites has been interpreted by means of crystal structures of HIV-1 protease with unhydrolysable peptide analogs (Mahalingam et al., 2001; Tie et al., 2005) or using the catalytically inactive D25N mutant with peptides (Prabu-Jeyabalan et al., 2000, 2002). The substrates or peptidic inhibitors bind in the active site cavity formed by residues from both subunits: 8, 23–30, 32, 45–50, 53, 56, 76, 80–82 and 84. Several of these residues also contribute to the dimer interface (especially Arg8, Ile50, Phe53), so that altered dimer formation will influence substrate binding. Peptides are bound within the active site cavity by means of a series of conserved hydrogen bond interactions as well as hydrophobic contacts of peptide side chains within subsites formed by protease residues. The substrate peptide bound within the protease subsites is illustrated schematically in Fig. 2.3, and the protease-peptide hydrogen bond interactions are shown in Fig. 2.4. One notable feature is the beta-sheet-like hydrogen bonds formed between main chain carbonyl oxygens and amides of the peptide from P3 to P3′ and the protease, which are conserved in the majority of protease-inhibitor structures (Gustchina et al., 1994). A conserved water molecule forms bridging hydrogen bond interactions between the flaps, the peptide main chain and the protease active site. The S1 and S1′ subsites are large and hydrophobic, as expected from the preference for large hydrophobic residues like Phe at P1/P1′. In contrast, subsites S2 and S2′ are smaller with polar contributions from Asp29 and Asp30, consistent with the preferences for small hydrophobic Val/Ile at P2/P2′ or polar Asn/Gln. The S4/4′ and S3/3′ subsites are partially exposed on the protease surface and can accommodate a greater variety of side chains. The hydrogen

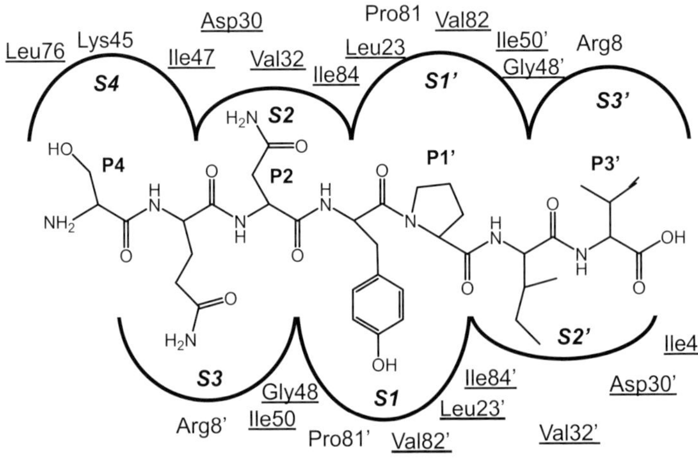

Fig. 2.3 Scheme of peptide substrate bound in active site cavity of the HIV protease dimer. The peptide substrate Ser-Gln-Asn-Tyr-Pro-Ile-Val (P4-P3′), which represents the MA/CA cleavage site in the Gag-Pro-Pol precursor, is shown in subsites S4-S3′ formed by protease residues. The curved line indicates the approximate size of each subsite. The residues contributing to the subsites are labeled and the underlined residues are mutated in drug resistance

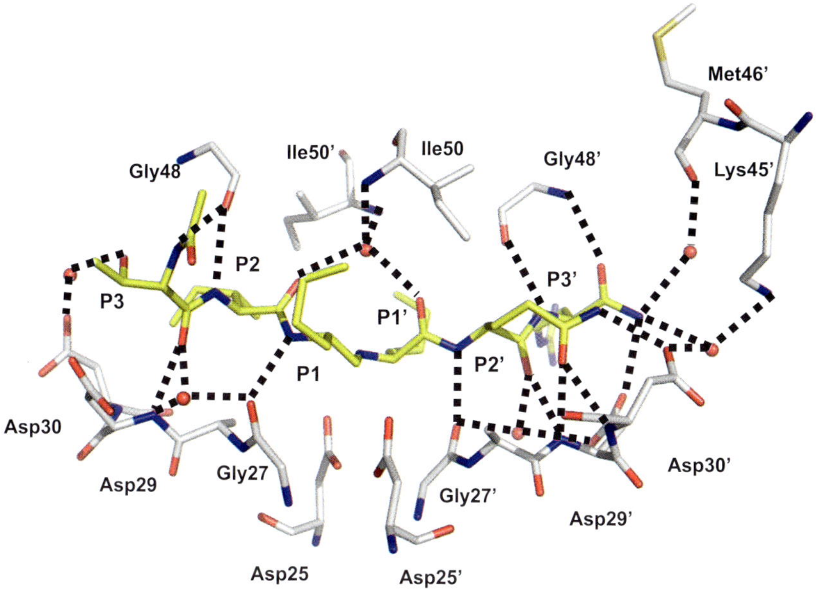

Fig. 2.4 Hydrogen bond interactions of HIV protease with a non-hydrolysable peptide analog. The non-hydrolysable peptide analog Acetyl-Thr-Ile-Nle-r-Nle-Gln-Arg-CONH2, where r is the reduced peptide (CH2–NH), is shown in yellow bonds with the interacting protease residues in grey bonds from the crystal structure (Tie et al., 2005). Asp25 and Asp25′ are the catalytic residues. Water molecules are shown as red spheres. Hydrogen bond interactions are indicated by dotted lines

bonds observed between the peptide analogs and the conserved regions of HIV protease have provided a template for the design of non-peptide inhibitors with equivalent polar interactions. Peptide analogs form at least seven hydrogen bonds with main chain protease atoms (Fig. 2.4), suggesting that equivalent interactions can be incorporated in the designs for non-peptide inhibitors.

2.3.3 Antiviral Inhibitors for HIV/AIDS Therapy

Many of the clinical inhibitors of HIV-1 protease were developed by means of structure-guided designs (Wlodawer and Vondrasak, 1998). Currently, nine protease inhibitors are used for AIDS therapy: amprenavir, atazanavir, darunavir, indinavir, lopinavir, nelfinavir, ritonavir, saquinavir and tipranavir. In current practice, ritonavir is used almost exclusively at subtherapeutic doses together with another HIV-1 protease inhibitor in order to inhibit the cytochrome P-450 CYP2A4 that degrades the drugs. These drugs all act by binding in the active site cavity of the HIV protease. The first antiviral protease inhibitors, saquinavir, indinavir and ritonavir, were designed to target the wild type HIV protease by mimicking the binding of substrates. These inhibitors contain a central hydroxyl group between the P1 and P1′ side groups that binds to the catalytic aspartates and mimics one hydroxyl of the tetrahedral reaction intermediate. The inhibitors bind to the protease by means of hydrogen bond interactions similar to those of peptides, and large hydrophobic side groups that fit in the hydrophobic protease subsites from S2 to S2′ and increase the binding affinity. The initial drugs retain several peptidic groups. Subsequently, inhibitors were designed with reduced peptidic characteristics, and the most recent inhibitor designs for tipranavir and darunavir have explicitly targeted drug resistant mutants of the enzyme, as described in the later sections.

2.4 The Challenge of Drug Resistance

2.4.1 Resistant Mutations in HIV-1 Protease

The HIV-1 protease sequence is naturally variable due to the lack of proofreading function of the HIV reverse transcriptase and consequent accumulation of sequence errors. This high mutational rate of the virus encourages the evolution of resistance to drugs. The emergence of resistance to protease inhibitors is associated with the appearance of amino acid substitutions in the viral protease gene that are rarely observed as natural polymorphisms (Shafer and Schapiro, 2008). Additionally, mutations in some of the protease cleavage sites have been found in resistance. Protease mutants with even minor alterations in activity will effect viral replication. Virus bearing protease with drug resistant mutations shows defects in polyprotein processing and reduced infectivity in several studies (Zennou et al., 1998; Martinez-Picado et al., 1999; Resch et al., 2002; Watkins et al., 2003). It is likely that primary mutations producing lower protease activity or stability, as well as reduced susceptibility to the drug, combine with compensating mutations to produce higher protease activity and viral replication in resistant HIV (Menéndez-Arias et al., 2003; van Maarseveen et al., 2007). To date, among the 99 amino acid residues of the protease, mutations of nearly 50 residues have been implicated in drug resistance. The most critical mutations that reduce viral susceptibility to different drugs are illustrated in Fig. 2.3. For some drugs, one critical mutation is sufficient to reduce viral susceptibility. Individualized therapy has been proposed for resistant HIV based on knowledge of genotypic data for mutations in the protease gene (Shafer et al., 2007).

2.4.2 Molecular Mechanisms of Drug Resistance

Analysis of protease mutants has revealed diverse mechanisms of resistance. The effects of drug resistance mutations have been dissected by examining HIV protease with single mutations, as described in (Louis et al., 2007; Weber et al., 2007).

The structural locations of drug-resistant mutations vary from substrate/inhibitor binding site, dimer interface, and flap region to the protease surface (Fig. 2.5). Mutations can directly, or indirectly, alter the protease catalytic activity, inhibitor binding, or the dimer stability, leading to drug resistance. Different categories of resistance mutations have been proposed based on their structural locations.

(a) Mutations of residues forming the active site cavity can directly interfere with the binding of the inhibitor. This category includes resistance mutations of D30N, V32I, G48V, I50V, V82A and I84V. These mutations generally resulted in loss of specific protease-inhibitor interactions (Mahalingam et al., 2004; Tie et al., 2004; Liu et al., 2005, 2008; Kovalevsky et al., 2006a, b), although the V82A mutant adapted by shifts of the backbone to reform favorable interactions with inhibitors. These mutations may also show reduced catalytic activity.
(b) Mutations that alter the dimer interface show reduced dimer stability, such as I50V, L24I and F53L (Liu et al., 2005, 2006). Mutations that disrupt the protease dimer are presumed to cause release of the drug without substantially reducing activity on the viral polyproteins so that infectious virus can be produced. The conformation and stability of the dimer are sensitive to mutations that alter intersubunit interactions in the presence (Liu et al., 2005) or absence of inhibitor (Liu et al., 2006). Some mutations like I50V fall into both category (a) and (b).
(c) Mutation of residues outside the active site cavity or dimer interface can indirectly alter protease activity, inhibition or stability. Resistance mutations in this category include G73S, L76V, N88D and L90M. Mutation L90M alters

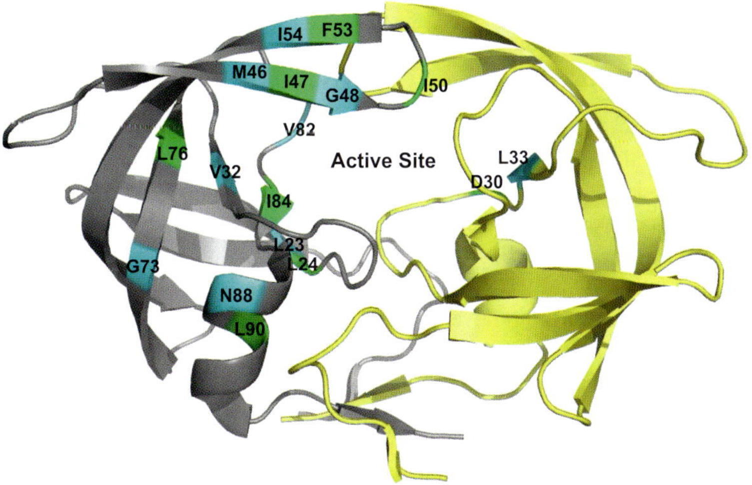

Fig. 2.5 Location of drug resistance mutations. The protease dimer is shown in ribbon representation with subunits colored yellow and grey. The sites of non-polymorphic residues that are mutated in drug resistant HIV are indicated in green and cyan. Many mutations alter residues forming the active site cavity or flexible flaps. Other mutations alter distal residues

interactions with the main chain carbonyl oxygen of the catalytic Asp25 (Mahalingam et al., 2001; Kovalevsky et al., 2006a). Mutation G73S showed variations in activity on different substrates relative to wild type enzyme by indirect influences via long-range structural perturbations (Liu et al., 2005), similar to the varying activity observed for D30N (Mahalingam et al., 1999).

(d) The protease structure and activity are sensitive to mutations in the flexible flaps. Mutations of several flap residues are observed in drug resistance including M46L, I47V, G48V, I50V, F53L, and I54V. Some mutants show different flap conformations in the absence or presence of substrate. A novel mechanism for resistance was suggested by the structure of unliganded mutant F53L, which altered the interactions between the two flaps in the open conformation (Liu et al., 2006). Another distinct mechanism of drug resistance was revealed by the structure of the I54V mutant with the tetrahedral reaction intermediate, which had lost water-mediated hydrogen bonds with the flap residues Ile50 and Ile50′ (Kovalevsky et al., 2007).

In summary, resistant mutants show changes in catalytic rate, specificity for cleavage sites and formation of stable dimers (reviewed in Louis et al., 2000, 2007; Weber et al., 2007), which will influence correct polyprotein processing and viral replication. The effects of mutations of the residues forming the active site cavity and interacting directly with the inhibitors are relatively easy to understand. Many resistant mutations alter residues at distal locations, which are more difficult to understand without detailed structural and kinetic measurements. The characterization of the structure and activity of resistant variants of the protease can provide valuable information to guide the selection of drugs for salvage therapy and the design of new inhibitors for resistant HIV.

2.5 Comparative Analysis of Retroviral Proteases

Various retroviral proteases have been studied, typically in comparison with HIV-1 protease. HIV-1 and other retroviral proteases share conserved sequences and structures (Wlodawer and Gustchina, 2000). Similar to HIV-1 protease, the human T-cell lymphotropic virus type-1 (HTLV-1) protease is a potential target for chemotherapy, as approximately 10–20 million people are infected worldwide by this virus, and about 5% of them develop disease (Bangham, 2000). The structure of a truncated HTLV-1 protease was determined only recently (Li et al., 2005). HTLV-1 protease has been reviewed recently (Tözsér and Weber, 2007). As exemplified by HTLV-1 protease (Kádas et al., 2004), clinical inhibitors of HIV-1 protease typically only weakly inhibit, if at all, the other retroviral proteases. Therefore, novel inhibitors must be designed for the proteases of other retroviruses associated with diseases. Development of such inhibitors will benefit greatly from the extensive efforts in structure-guided drug design for HIV protease.

Comparative analysis of the proteases of many retroviruses has helped to define the residues that are critical for catalytic activity and substrate specificity. Studies of members of each retroviral genus with a large panel of substrates have revealed distinct substrate preferences for HIV-1 protease (Bagossi et al., 2005; Eizert et al., 2008). Studies on diverse retroviral proteases will help design broad-spectrum inhibitors for many retroviral infections. Broad-spectrum inhibitors are expected to retain their effectiveness on resistant mutants of HIV-1 as well as for diverse subtypes of HIV-1 and HIV-2 (Menéndez-Arias and Tözsér, 2008).

2.6 Structure-Guided Drug Design: Current and Future Prospects

2.6.1 Structure-Guided Design of Darunavir for Resistant HIV

One of the most exciting advances in recent years has been the development of antiviral drugs that target resistant HIV. Our investigations have focused on darunavir (originally named TMC114 and UIC-94017), which was approved in June 2006 for salvage therapy of AIDS patients who failed other treatments. The design goal for darunavir was to introduce new interactions with the HIV protease main chain in order to reduce the deleterious effects of mutations (Ghosh et al., 2008a). Achievement of this goal was verified by our crystal structures and antiviral data (Koh et al., 2003; Tie et al., 2004) and analyses of others (King et al., 2004; Surleraux et al., 2005). The crystal structures demonstrated new hydrogen bonds of darunavir with the main chain atoms of Asp 29 and 30 in the protease (Tie et al., 2004), which are absent in the complex with the chemically related amprenavir (Kim et al., 1995). The darunavir interactions with HIV-1 protease are illustrated in Fig. 2.6. These new interactions with main chain atoms were proposed to be critical for the potency and broad-spectrum activity on multi-drug resistant strains of HIV. Darunavir binds with higher affinity, slower dissociation and favorable enthalpy relative to other inhibitors (King et al., 2004; Dierynck et al., 2007), and blocks formation of protease dimers (Koh et al., 2007). It has been observed bound at a second site on the flap of the protease (Kovalevsky et al., 2006), and kinetic analysis has showed mixed-type competitive-uncompetitive inhibition for darunavir (Kovalevsky et al., 2008b). Most important, darunavir is extremely potent against many HIV-1 strains and resistant clinical isolates (IC_{50} of 3 nM) with minimal cytotoxicity (Koh et al., 2003). The effects of darunavir on the structures of various protease mutants are summarized in (Weber et al., 2007). Our analysis of the structural and kinetic effects of darunavir on protease with single mutations of D30N, V32I, M46L, I50V, I54V, I54M, V82A, I84V and L90M is described in (Tie et al., 2004; Kovalevsky et al., 2006a, b; Liu et al., 2008).

In the clinic, darunavir has performed exceptionally well, showing excellent virological responses and favorable safety and tolerance in treatment-experienced patients (Clotet et al., 2007; Rachlis et al., 2007). It was more effective in salvage

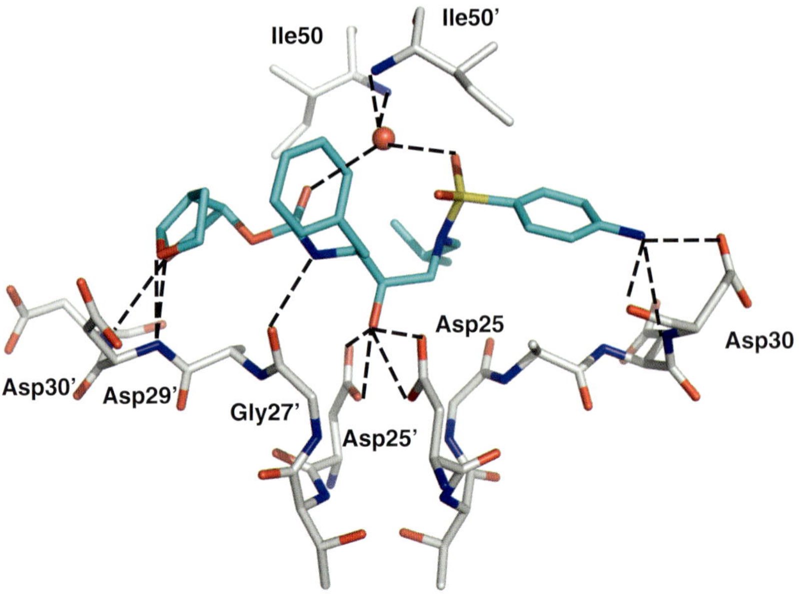

Fig. 2.6 Darunavir interactions with HIV-1 protease. The clinical inhibitor darunavir is shown in blue bonds with the interacting HIV protease residues in grey bonds. The conserved water molecule is shown as a red sphere. Hydrogen bond interactions are indicated by dotted lines

therapy than other tested clinical inhibitors, and showed lower toxicity than tipranavir (Temesgen and Feinberg 2007). Darunavir retains potency on non-B subtypes of HIV-1 and on HIV-2 infections (Poveda et al., 2008; Desbois et al., 2008). Importantly, darunavir shows a very high genetic barrier to resistance. Diminished clinical efficacy appeared when isolates had ten-fold decreased susceptibility (deMeyer et al., 2008), while 10% decreased susceptibility is sufficient to eliminate clinical activity of the other tested protease inhibitors. At least three of protease mutations V11I, V32I, L33F, I47V, I50V, I54L/M, G73S, L76V, I84V and L89V were required to produce diminished clinical response to daruanvir (de Meyer et al., 2008). Analysis of these multi-drug resistant (MDR), multiply mutated protease variants observed in clinical isolates with reduced susceptibility to darunavir will help to improve the next inhibitor designs.

2.6.2 New Inhibitor Designs

The severe problem of drug resistance means it is critical to develop new inhibitors that target resistant HIV. The development of new inhibitor designs has been guided by structural and inhibition data, especially for drug resistant protease variants. Several recent studies are described. All the drugs and the majority of successful antiviral inhibitors target the active site cavity of HIV protease. Potential alternate protease sites are being investigated, as described later. The most common, and frequently effective, design strategy is focused on variations of known antiviral inhibitors. Subnanomolar inhibitors have been designed in this structure-guided strategy by targeting the substrate-binding cavity of the protease (Reddy et al., 2007; Altman et al., 2008). Our strategy has been to design inhibitors incorporating new hydrogen bond interactions with conserved regions of the protease active site cavity (Ghosh et al., 2008a). This strategy has produced a series of new antiviral inhibitors designed with various P2 and P2′ groups on the chemical framework of darunavir with the aim of introducing distinct polar interactions with the protease main chain. The designed compounds were verified to inhibit HIV-1 protease and the mode of inhibition determined by analysis of crystal structures of the inhibitor complex with HIV-1 protease. The design strategy is exemplified by a detailed series of chemical derivatives of the P2 group leading to the antiviral inhibitor GRL-0255A, which showed picomolar inhibition of the protease and nanomolar antiviral activity on cells infected with HIV-1, HIV-2 and multiple drug resistant variants of HIV-1 (Ghosh et al., 2008b). Several of the best antiviral inhibitors are shown in Fig. 2.7, including GRL-06579A (Ghosh et al., 2006), GRL-98065 (Amano et al., 2007) and GRL-0255A (Ghosh et al., 2008b). Like darunavir, the antiviral inhibitor GRL-98065 showed minimal changes in the crystal structures and interactions with resistant mutants of HIV protease (Wang et al., 2007). In fact minor structural changes were observed for darunavir, GRL-06579A and GRL-98065 complexes, even with HIV-2 protease, which differs in about 50/99 residues from HIV-1 protease (Kovalevsky et al., 2008a). Importantly, these inhibitors are expected to maintain their potency on resistant HIV due to the retention of antiviral efficacy on HIV variants. The best of these antiviral inhibitors will move into further preclinical studies as potential drugs for AIDS.

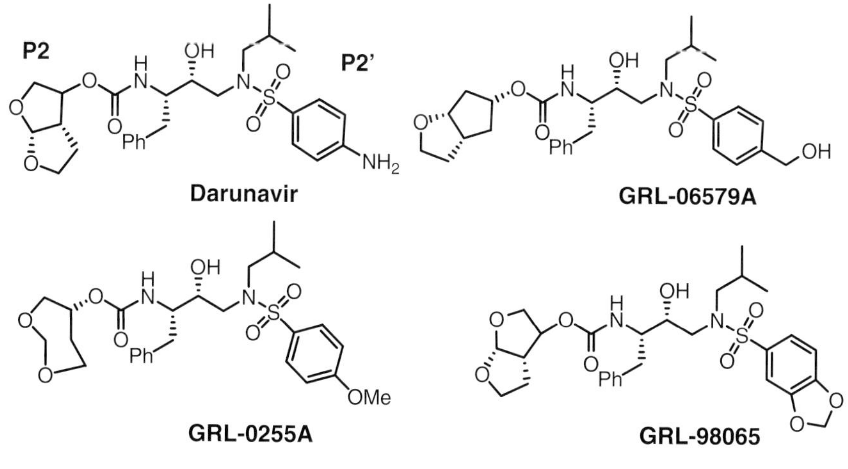

Fig. 2.7 Inhibitors based on the darunavir chemical scaffold. Darunavir and other antiviral inhibitors are shown. The inhibitors differ in the groups at P2 and P2′

Another possibility for designing protease inhibitors with a distinct mechanism of action is based on new crystal structures with reaction intermediates. Recent studies have shed light on the protease reaction mechanism by illustrating the binding of tetrahedral reaction intermediates (Kumar et al., 2005; Kovalevsky et al., 2007) or product complexes (Tyndall et al., 2008). Importantly, these structures will define the catalytic geometry and shed light on the reaction mechanism. Quantum calculations based on these structures will assist in the design of novel mechanism-based inhibitors. Ideally, these inhibitors will target the active site by incorporating a distinct chemistry, unlike the standard hydroxyl group interacting with the catalytic Aspartates that is used in all the current drugs.

2.6.3 Alternate Targets on Protease for Inhibitor Design

Various targets on the protease have been investigated beyond the standard active site cavity where substrates bind. These alternate targets include the dimer interface, the open conformation of the protease, and the surface of the flexible flaps. New studies attacking these targets are described, although none of these inhibitors has yet moved into clinical trials.

A number of studies have targeted the dimer interface of the protease with the goal of preventing formation of the active protease dimer (Camarasa et al., 2006). The initial approach was to use peptides to mimic one of the termini forming the four-stranded beta sheet, which is an important component of the dimer interface (Weber, 1990). Peptides can be linked to improve their potency, as described in (Bowman and Chmielewski, 2008). Several peptides and other types of molecules have been shown to bind to the dimer interface and inhibit the protease activity (Bannwarth and Reboud-Ravaux, 2007). Moreover, some active site inhibitors like darunavir, also have shown inhibition of dimer formation (Koh et al., 2008). An alternate approach is to design macromolecular inhibitors consisting of mutated inactive protease, which formed heterodimers with active protease subunits and showed reduced viral infectivity (Miklossy et al., 2008). Also monoclonal antibodies have been designed to bind protease with nanomolar inhibition of protease activity and inhibition of dimer formation (Bartonová et al., 2008). These antibodies were also active on resistance mutants of HIV protease.

Targets on the molecular surface of the protease, particularly in the flaps, have been studied. Some reversible inhibitor molecules, like beta-lactam compounds and Nb-containing polyoxometalates, were proposed to bind on the surface of the enzyme rather than in the usual active site cavity (Sperka et al., 2005; Judd et al., 2001). Peptide and haloperidol-based irreversible inhibitors were observed bound on the protease surface (Brynda et al., 2004; DeVoss et al., 1994). Novel inhibitors, albeit of modest micromolar inhibition, have been discovered in virtual screens of compounds targeting the tip of the flap (Damm et al., 2008). Other micromolar inhibitors have been developed, and confirmed, to target the open flap conformation (Böttcher et al., 2008). Novel inorganic metallacarborane inhibitors have been developed with the similar target of the open flap conformation and crystallographic verification of the binding mode (Cigler et al., 2005; Kozisek et al., 2008). The crystal structure of the inhibitor 3-cobalt bis(1,2-dicarbollide) bound in the protease is shown in Fig. 2.8. Two inhibitor molecules bind between the flaps and residues 81–84. Nanomolar inhibitors have been derived from this compound suggesting the potential of boron-based inhibitors for future therapeutic development.

We have discovered a novel surface binding site for darunavir in the highest resolution structure to date (0.84 Å resolution) for the HIV-1 protease V32I mutant (Kovalevsky et al., 2006b). This sub-atomic resolution structure was remarkable in several aspects. Two conformers of the protease dimer were observed: one protease dimer bound darunavir in the active site, the other dimer bound darunavir at two sites, the normal active site cavity and a distinct site on the surface of one flap. Remarkably, darunavir had different diastereomers, related by nitrogen invertion, bound at the active site and the second flap site. This second binding site provides a novel target for drug design. Our subsequent kinetic analysis showed mixed-type competitive-uncompetitive inhibition for darunavir and amprenavir, unlike the purely competitive inhibition of saquinavir (Kovalevsky et al., 2008b). Therefore, new compounds designed to bind tightly to this flap site have potential for development as antiviral agents with a distinct allosteric mechanism.

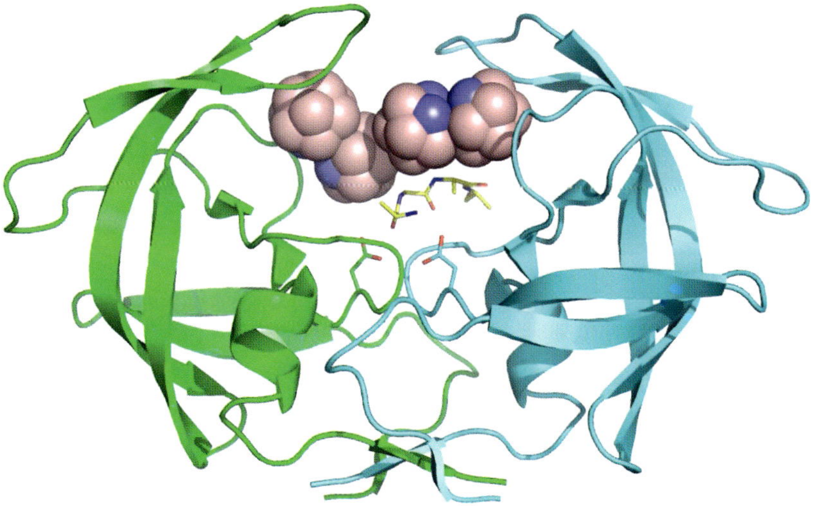

Fig. 2.8 Metallacarborane inhibitor bound to HIV protease. The inhibitor 3-cobalt bis(1,2-dicarbollide) is shown as pink (boron) and purple (carbon) space-filling atoms. The cobalt atoms are buried within the cages of boron and carbon atoms. Two inhibitor molecules bind in the HIV protease dimer, which is colored green and cyan for the two subunits. A bound peptide Ala-Gly-Ala-Ala is in yellow bonds

2.7 Conclusion

Antiviral inhibitors of HIV-1 protease have provided a decade of successful therapy for HIV/AIDS, which dramatically increased the lifespan of infected people since the first drug targeting the protease was used in 1995. Detailed atomic structures and binding data for HIV protease mutants and the inhibitors continue to be important in the design of future drugs. Novel approaches like targeting the dimer interface, the open flaps, or potential allosteric sites are being studied. A variety of promising inhibitors have resulted, which provide leads for optimization and antiviral assays. At present, however, the most powerful antiviral inhibitors act by directly targeting the active site of the enzyme. In recent years, the strategy has focused on attacking drug resistant HIV. New inhibitors, such as darunavir, designed on structural principles to target resistant variants of the protease, are very potent drugs. This strategy of designing inhibitors to form more strong polar interactions with backbone and conserved protease residues has been successful in combating drug resistance. Several new antiviral inhibitors have been designed on the same principles. The structure-guided strategies hold promise for future AIDS therapies. Also, knowledge of the mechanisms of drug resistance has broader implications for treatment of numerous diseases where drug resistance is a critical health issue, such as bacterial infections, tuberculosis, and cancer.

Acknowledgements This research was supported in part by the Georgia State University Molecular Basis of Disease Program, the Georgia Research Alliance, the Georgia Cancer Coalition, the United States National Institutes of Health and the Hungarian Science and Research Fund (OTKA K68288).

References

Altman, M.D., Ali, A., Reddy, G.S., Nalam, M.N., Anjum, S.G., Cao, H., Chellappan, S., Kairys, V., Fernandes, M.X., Gilson, M.K., Schiffer, C.A., Rana, T.M., Tidor, B. 2008, HIV-1 protease inhibitors from inverse design in the substrate envelope exhibit subnanomolar binding to drug-resistant variants. *J Am Chem Soc.* **130**: 6099–6113.

Amano, M., Koh, Y., Das, D., Wang, Y.-F., Boross, P.I., Li, J., Leschenko, S., Weber, I.T., Ghosh, A.K., Mitsuya, H. 2007, A Novel *bis*-tetrahydrofuranylurethane-containing nonpeptidic protease inhibitor (PI) GRL-98065 potent against multi-PI-resistant HIV *in vitro*. *Antimicrob Agents Chemother.* **51**: 2143–2155.

Bagossi, P., Sperka, T., Fehér, A., Kádas, J., Zahuczky, G., Miklóssy, G., Boross, P., Tözsér, J. 2005, Amino acid preferences for a critical substrate binding subsite of retroviral proteases. *J Virol.* **79**: 4213–4218.

Bangham, C.R. 2000, HTLV-1 infections. *J Clin Pathol.* **53**: 581–586.

Bannwarth, L., Reboud-Ravaux, M. 2007, An alternative strategy for inhibiting multidrug-resistant mutants of the dimeric HIV-1 protease by targeting the subunit interface. *Biochem Soc Trans.* **35**: 551–554.

Bartonová, V., Král, V., Sieglová, I., Brynda, J., Fábry, M., Horejsí, M., Kozísek, M., Sasková, K.G., Konvalinka, J., Sedlácek, J., Rezácová, P. 2008, Potent inhibition of drug-resistant HIV protease variants by monoclonal antibodies. *Antiviral Res.* **78**: 275–277.

Beck, Z.Q., Morris, G.M., Elder, J.H. 2002, Defining HIV-1 protease substrate specificity. *Curr Drug Targets Infect Disord.* **2**: 37–50.

Böttcher, J., Blum, A., Dörr, S., Heine, A., Diederich, W.E., Klebe, G. 2008, Targeting the open-flap conformation of HIV-1 protease with Pyrrolidine-based inhibitors. *Chem Med Chem.* **3**: 1337–1344.

Bowman, M.J., Chmielewski, J. 2009, Sidechain-linked inhibitors of HIV-1 protease dimerization. *Bioorg Med Chem.* **17**: 967–976.

Brynda, J., Rezacova, P., Fabry, M., Horejsi, M., Stouracova, R., Soucek, M., Hradilek, M., Konvalinka, J., Sedlacek, J. 2004, Inhibitor binding at the protein interface in crystals of a HIV-1 protease complex. *Acta Crystallogr.* **D60**: 1943–1948.

Buonaguro, L., Tornesello, M.L., Buonaguro, F.M. 2007, Human immunodeficiency virus type 1 subtype distribution in the worldwide epidemic: pathogenetic and therapeutic implications. *J Virol.* **81:** 10209–10219.

Camarasa, M.J., Velázquez, S., San-Félix, A., Pérez-Pérez, M.J., Gago, F. 2006, Dimerization inhibitors of HIV-1 reverse transcriptase, protease and integrase: a single mode of inhibition for the three HIV enzymes? *Antiviral Res.* **71**: 260–267.

Cígler, P., Kozísek, M., Rezácová, P., Brynda, J., Otwinowski, Z., Pokorná, J., Plesek, J., Grüner, B., Dolecková-Maresová, L., Mása, M., Sedlácek, J., Bodem, J., Kräusslich, H.G., Král, V., Konvalinka, J. 2005, From nonpeptide toward noncarbon protease inhibitors: metallacarboranes as specific and potent inhibitors of HIV protease. *Proc Natl Acad Sci USA.* **102**: 15394–15399.

Clotet, B., Bellos, N., Molina, J.M., Cooper, D., Goffard, J.C., Lazzarin, A., Wöhrmann, A., Katlama, C., Wilkin, T., Haubrich, R., Cohen, C., Farthing, C., Jayaweera, D., Markowitz, M., Ruane, P., Spinosa-Guzman, S., Lefebvre, E., POWER 1 and 2 study groups. 2007, Efficacy and safety of darunavir-ritonavir at week 48 in treatment-experienced patients with HIV-1 infection in POWER 1 and 2: a pooled subgroup analysis of data from two randomised trials. *Lancet.* **369**: 1169–1178.

Damm, K.L., Ung, P.M., Quintero, J.J., Gestwicki, J.E., Carlson, H.A. 2008, A poke in the eye: inhibiting HIV-1 protease through its flap-recognition pocket. *Biopolymers.* **89**: 643–652.

De Meyer, S., Vangeneugden, T., van Baelen, B., de Paepe, E., van Marck, H., Picchio, G., Lefebvre, E., de Béthune, M.P. 2008, Resistance profile of darunavir: combined 24-week results from the POWER trials. *AIDS Res Hum Retroviruses.* **24**: 379–388.

Desbois, D., Roquebert, B., Peytavin, G., Damond, F., Collin, G., Bénard, A., Campa, P., Matheron, S., Chêne, G., Brun-Vézinet, F., Descamps, D., for the French ANRS HIV-2 Cohort (ANRS CO 05 VIH-2). 2008, In vitro phenotypic susceptibility of human immunodeficiency virus type 2 clinical isolates to protease inhibitors. *Antimicrob Agents Chemother.* **52**: 1545–1548.

De Voss, J.J., Sui, Z., DeCamp, D.L., Salto, R., Babe, L.M., Craik, C.S., Ortiz de Montellano, P.R. 1994, Haloperidol-based irreversible inhibitors of the HIV-1 and HIV-2 proteases. *J Med Chem.* **37**: 665–673.

Dierynck, I., De Wit, M., Gustin, E., Keuleers, I., Vandersmissen, J., Hallenberger, S., Hertogs, K. 2007, Binding kinetics of darunavir to human immunodeficiency virus type 1 protease explain the potent antiviral activity and high genetic barrier. *J Virol.* **81**: 13845–13851.

Eizert, H., Bagossi, P., Sperka, T., Miklóssy, G., Bander, P., Boross, P., Weber, I.T., Tözsér, J. 2008, Amino acid preferences of retroviral proteases for amino-terminal positions in a type-1 cleavage site. *J Virol.* **82**: 10111–10117.

Ghosh, A.K., Sridhar, P.R., Leshchenko, S., Hussain, A.K., Li, J., Kovalevsky, A.Y., Walters, D.E., Wedekind, J.E., Tokars, V.L., Das, D., Koh, Y., Maeda, K., Gatanaga, H., Weber, I.T., Mitsuya, H. 2006, Structure-based design of novel HIV-1 protease inhibitors to combat drug resistance. *J Med Chem.* **49**: 5252–5261.

Ghosh, A.K., Chapsal, B.D., Weber, I.T., Mitsuya, H. 2008a, Design of HIV protease inhibitors targeting protein backbone: an effective strategy for combating drug resistance. *Acc Chem Res.* **41**: 78–86.

Ghosh, A.K., Gemma, S., Baldridge, A., Leschenko, S., Wang, Y.F., Kovalevsky, A.Y., Koh, Y., Weber, I.T., Mitsuya, H. 2008b, Flexible cyclic ethers/polyethers as novel P2-ligands for HIV-1 protease inhibitors: design, synthesis, biological evaluation and protein-ligand X-ray studies. *J Med Chem.* **51**: 6021–6033.

Gustchina, A., Sansom, C., Prevost, M., Richelle, J., Wodak, S.Y., Wlodawer, A., Weber, I.T. 1994, Energy calculations and analysis of HIV-1 protease-inhibitor crystal structures. *Protein Eng.* **7**: 309–317.

Hammer, S.M., Eron, J.J. Jr., Reiss, P., Schooley, R.T., Thompson, M.A., Walmsley, S., Cahn, P., Fischl, M.A., Gatell, J.M., Hirsch, M.S., Jacobsen, D.M., Montaner, J.S., Richman, D.D., Yeni, P.G., Volberding, P.A., International AIDS Society-USA. 2008, Antiretroviral treatment of adult HIV infection: 2008 recommendations of the International AIDS Society-USA panel. *JAMA.* **300**: 555–570.

Judd, D.A., Nettles, J.H., Nevis, N., Snyder, J.P., Liotta, D.C., Tang, J., Ermolieff, J., Schinazi, R.F., Hill, C.L. 2001, Polyoxometalate HIV-1 protease inhibitors. A new mode of protease inhibition. *J Am Chem Soc.* **123**: 886–897.

Kádas, J., Weber, I.T., Bagossi, P., Miklóssy, G., Boross, P., Oroszlan, S., Tözsér, J. 2004, Narrow substrate specificity and sensitivity toward ligand-binding site mutations of human T-cell Leukemia virus type 1 protease. *J Biol Chem.* **279**: 27148–27157.

Kim, E.E., Baker, C.T., Dwyer, M.D., Murcko, M.A., Rao, B.G., Tung, R.D., Navia, M.A. 1995, Crystal structure of HIV-1 protease in complex with VX-478, a potent and orally bioavailable inhibitor of the enzyme. *J Am Chem Soc.* **117**: 1181–1182.

King, N.M., Prabu-Jeyabalan, M., Nalivaika, E.A., Wigerinck, P., de Béthune, M.P., Schiffer, C.A. 2004, Structural and thermodynamic basis for the binding of TMC114, a next-generation human immunodeficiency virus type 1 protease inhibitor. *J Virol.* **78**: 12012–12021.

Koh, Y., Nakata, H., Maeda, K., Ogata, H., Bilcer, G., Devasamundam, T., Kincaid, J.F., Boross, P., Wang, Y.F., Tie, Y., Volarath, P., Gaddis, L., Harrison, R.W., Weber, I.T., Ghosh, A.K., Mitsuya, H. 2003, Novel bis-tetrahydrofuranylurethane-containing nonpeptidic protease inhibitor (PI) UIC-94017 (TMC114) potent against multi-PI-resistant HIV in vitro. *Antimicrob Agent Chemother.* **47**: 3123–3129.

Koh, Y., Matsumi, S., Das, D., Amano, M., Davis, D.A., Li, J., Leschenko, S., Baldridge, A., Shioda, T., Yarchoan, R., Ghosh, A.K., Mitsuya, H. 2007, Potent inhibition of HIV-1 replication by novel non-peptidyl small molecule inhibitors of protease dimerization. *J Biol Chem.* **282**: 28709–28720.

Kovalevsky, A.Y., Tie, Y., Liu, F., Boross, P.I., Wang, Y.F., Leshchenko, S., Ghosh, A.K., Harrison R.W., Weber I.T., 2006a, Effectiveness of nonpeptide clinical inhibitor TMC114 on HIV-1 protease with highly drug resistant mutations D30N, I50V and L90M. *J Med Chem.* **49**: 1379–1387.

Kovalevsky, A.Y., Liu, F., Leshchenko, S., Ghosh, A.K., Louis, J.M., Harrison, R.W., Weber, I.T. 2006b, Ultra-high resolution crystal structure of HIV-1 protease mutant reveals two binding sites for clinical inhibitor TMC114. *J Mol Biol.* **363**: 161–173.

Kovalevsky, A.Y., Chumanevich, A.A., Liu, F., Weber, I.T. 2007, Caught in the act: 1.5 Å resolution crystal structures of the HIV-1 protease and the I54V mutant reveal a tetrahedral reaction intermediate. *Biochemistry.* **46**: 14854–14864.

Kovalevsky, A.Y., Louis, J.M., Aniana, A., Ghosh, A.K., Weber, I.T. 2008a, Structural evidence for effectiveness of darunavir and two related antiviral inhibitors against HIV-2 protease. *J Mol Biol.* **384**: 178–192.

Kovalevsky, A.Y., Ghosh, A.K., Weber, I.T. 2008b, Solution kinetics measurements suggest HIV-1 protease has two binding sites for darunavir and amprenavir. *J Med Chem.* **51**: 6599–6603.

Kozísek, M., Cígler, P., Lepsík, M., Fanfrlík, J., Rezácová, P., Brynda, J., Pokorná, J., Plesek, J., Grüner, B., Grantz Sasková, K., Václavíková, J., Král, V., Konvalinka, J. 2008, Inorganic polyhedral metallacarborane inhibitors of HIV protease: a new approach to overcoming antiviral resistance. *J Med Chem.* **51**: 4839–4843.

Kumar, M., Prashar, V., Mahale, S., Hosur, M.V. 2005, Observation of a tetrahedral reaction intermediate in the HIV-1 protease-substrate complex. *Biochem J.* **389**: 365–371.

Lagnese, M., Daar, E.S. 2008, Antiretroviral regimens for treatment-experienced patients with HIV-1 infection. *Expert Opin Pharmacother.* **9**: 687–700.

Li, F., Goila-Gaur, R., Salzwedel, K., Kilgore, N.R., Reddick, M., Matallana, C., Castillo, A., Zoumplis, D., Martin, D.E., Orenstein, J.M., Allaway, G.P., Freed, E.O., Wild, C.T. 2003, PA-457: a potent HIV inhibitor that disrupts core condensation by targeting a late step in Gag processing. *Proc Natl Acad Sci USA*. **100**: 13555–13560.

Li, M., Laco, G.S., Jaskolski, M., Rozycki, J., Alexandratos, J., Wlodawer, A., Gustchina, A. 2005, Crystal structure of human T cell leukemia virus protease, a novel target for anticancer drug design. *Proc Natl Acad Sci USA*. **102**: 18332–18337.

Liu, F., Boross, P.I., Wang, Y.F., Tözsér, J., Louis, J.M., Harrison, R.W., Weber, I.T. 2005, Kinetic, stability, and structural changes in high-resolution crystal structures of HIV-1 protease with drug-resistant mutations L24I, I50V, and G73S. *J Mol Biol*. **354**: 789–800.

Liu, F., Kovalevsky, A.Y., Louis, J.M., Boross, P.I., Wang, Y.F., Harrison, R.W., Weber, I.T. 2006, Mechanism of drug resistance revealed by the crystal structure of the unliganded HIV-1 protease with F53L mutation. *J Mol Biol*. **358**: 1191–1199.

Liu, F., Kovalevsky, A.Y., Tie, Y., Ghosh, A.K., Harrison, R.W., Weber, I.T. 2008, Effect of flap mutations on structure of HIV-1 protease and inhibition by saquinavir and darunavir. *J Mol Biol*. **381**: 102–115.

Louis, J., Wondrak, E.M., Kimmel, A.R., Wingfield, P.T., Nashed, N.T. 1999, Proteolytic processing of HIV-1 protease precursor, kinetics and mechanism. *J Biol Chem*. **274**: 23437–23442.

Louis, J.M., Weber, I.T., Tözsér, J., Clore, G.M., Gronenborn, A.M. 2000, HIV-1 protease: maturation, enzyme specificity, and drug resistance. *Adv Pharmacol*. **49**: 111–146.

Louis, J.M., Ishima, R., Torchia, D.A., Weber, I.T. 2007, HIV-1 protease: structure, dynamics and inhibition. *Adv Pharmacol*. **55**: 261–298.

Mahalingam, B., Louis, J.M., Reed, C.C., Adomat, J.M., Krouse, J., Wang, Y.F., Harrison, R.W., Weber, I.T. 1999, Structural and kinetic characterization of drug resistant mutants of HIV-1 protease. *Eur J Biochem*. **263**: 238–245.

Mahalingam, B., Louis, J.M., Hung, J., Harrison, R.W., Weber, I.T. 2001, Structural implications of drug-resistant mutants of HIV-1 protease: high-resolution crystal structures of the mutant protease/substrate analogue complexes. *Prot Struct Funct Genet*. **43**: 455–464.

Mahalingam, B., Wang, Y.-F., Boross, P.I., Tozser, J., Louis, J.M., Harrison, R.W., Weber, I.T. 2004, Crystal structures of HIV protease V82A and L90M mutants reveal changes in indinavir binding site. *Eur J Biochem*. **271**: 1516–1524.

Martinez-Cajas, J.L., Wainberg, M.A. 2007, Protease inhibitor resistance in HIV-infected patients: molecular and clinical perspectives. *Antiviral Res*. **76**: 203–221.

Martinez-Cajas, J.L., Wainberg, M.A. 2008, Antiretroviral therapy: optimal sequencing of therapy to avoid resistance. *Drugs*. **68**: 43–72.

Martinez Picado, J., Savara, A.V., Sutton, L., D'Aquila, R.T. 1999, Replicative fitness of protease inhibitor-resistant mutants of human immunodeficiency virus type 1. *J Virol*. **73**: 3744–3752.

Mastrolorenzo, A., Rusconi, S., Scozzafava, A., Barbaro, G., Supuran, C.T. 2007, Inhibitors of HIV-1 protease: current state of the art 10 years after their introduction. From antiretroviral drugs to antifungal, antibacterial and antitumor agents based on aspartic protease inhibitors. *Curr Med Chem*. **14**: 2734–2748.

Menéndez-Arias, L., Tözsér, J. 2008, HIV-1 protease inhibitors: effects on HIV-2 replication and resistance. *Trends Pharmacol Sci*. **29**: 42–49.

Menéndez-Arias, L., Martínez, M.A., Quiñones-Mateu, M.E., Martinez-Picado, J. 2003, Fitness variations and their impact on the evolution of antiretroviral drug resistance. *Curr Drug Targets Infect Disord*. **3**: 355–371.

Miklóssy, G., Tözsér, J., Kádas, J., Ishima, R., Louis, J.M., Bagossi, P. 2008, Novel macromolecular inhibitors of human immunodeficiency virus-1 protease. *Protein Eng Des Sel*. **21**: 453–461.

Pettit, S.C., Moody, M.D., Wehbie, R.S., Kaplan, A.H., Nantermet, P.V., Klein, C.A., Swanstrom, R. 1994, The p2 domain of human immunodeficiency virus type 1 Gag regulates sequential proteolytic processing and is required to produce fully infectious virions. *J Virol*. **68**: 8017–8027.

Pettit, S.C., Lindquist, J.N., Kaplan, A.H., Swanstrom, R. 2005, Processing sites in the human immunodeficiency virus type 1 (HIV-1) Gag-Pro-Pol precursor are cleaved by the viral protease at different rates. *Retrovirology*. **2**: 66–71.

Poveda, E., de Mendoza, C., Parkin, N., Choe, S., García-Gasco, P., Corral, A., Soriano, V. 2008, Evidence for different susceptibility to tipranavir and darunavir in patients infected with distinct HIV-1 subtypes. *AIDS*. **22**: 611–616.

Prabu-Jeyabalan, M., Nalivaika, E., Schiffer, C.A. 2000, How does a symmetric dimer recognize an asymmetric substrate? A substrate complex of HIV-1 protease. *J Mol Biol*. **301**: 1207–1220.

Prabu-Jeyabalan, M., Nalivaika, E., Schiffer, C.A. 2002, Substrate shape determines specificity of recognition for HIV-1 protease: analysis of crystal structures of six substrate complexes. *Structure*. **10**: 369–381.

Rachlis, A., Clotet, B., Baxter, J., Murphy, R., Lefebvre, E. 2007, Safety, tolerability, and efficacy of darunavir (TMC114) with low-dose ritonavir in treatment-experienced, hepatitis B or C co-infected patients in POWER 1 and 3. *HIV Clin Trials*. **8**: 213–220.

Reddy, G.S., Ali, A., Nalam, M.N., Anjum, S.G., Cao, H., Nathans, R.S., Schiffer, C.A., Rana, T.M. 2007, Design and synthesis of HIV-1 protease inhibitors incorporating oxazolidinones as P2/P2′ ligands in pseudosymmetric dipeptide isosteres. *J Med Chem*. **50**: 4316–4328.

Resch, W., Ziermann, R., Parkin, N., Gamarnik, A., Swanstrom, R. 2002, Nelfinavir-resistant, amprenavir-hypersusceptible strains of human immunodeficiency virus type 1 carrying an N88S mutation in protease have reduced infectivity, reduced replication capacity, and reduced fitness and process the Gag polyprotein precursor aberrantly. *J Virol*. **76**: 8659–8666.

Sayer, J.M., Liu, F., Ishima, R., Weber, I.T., Louis, J.M. 2008, Effect of the active site D25N mutation on the structure, stability, and ligand binding of the mature HIV-1 protease. *J Biol Chem*. **283**: 13459–13470.

Shafer, R.W., Schapiro, J.M. 2008, HIV-1 drug resistance mutations: an updated framework for the second decade of HAART. *AIDS Rev*. **10**: 67–84.

Shafer, R.W., Rhee, S.Y., Pillay, D., Miller, V., Sandstrom, P., Schapiro, J.M., Kuritzkes, D.R., Bennett, D. 2007, HIV-1 protease and reverse transcriptase mutations for drug resistance surveillance. *AIDS*. **21**: 215–223.

Sperka, T., Pitlik, J., Bagossi, P., Tözsér, J. 2005, Beta-lactam compounds as apparently uncompetitive inhibitors of HIV-1 protease. *Bioorg Med Chem Lett*. **15**: 3086–3090.

Surleraux, D.L., Tahri, A., Verschueren, W.G., Pille, G.M., de Kock, H.A., Jonckers, T.H., Peeters, A., De Meyer, S., Azijn, H., Pauwels, R., de Bethune, M.P., King, N.M., Prabu-Jeyabalan, M., Schiffer, C.A., Wigerinck, P.B. 2005, Discovery and selection of TMC114, a next generation HIV-1 protease inhibitor. *J Med Chem*. **48**: 1813–1822.

Temesgen, Z., Feinberg, J. 2007, Tipranavir: a new option for the treatment of drug-resistant HIV infection. *Clin Infect Dis*. **45**: 761–769.

Temesgen, Z., Feinberg, J.E. 2006, Drug evaluation: bevirimat – HIV Gag protein and viral maturation inhibitor. *Curr Opin Investig Drugs*. **7**: 759–765.

Tie, Y., Boross, P.I., Wang, Y.F., Gaddis, L., Hussain, A.K., Leshchenko, S., Ghosh, A.K., Louis, J.M., Harrison, R.W., Weber, I.T. 2004, High resolution crystal structures of HIV-1 protease with a potent non-peptide inhibitor (UIC-94017) active against multi-drug-resistant clinical strains. *J Mol Biol*. **338**: 341–352.

Tie, Y., Boross, P.I., Wang, Y.F., Gaddis, L., Liu, F., Chen, X., Tözsér, J., Harrison, R.W., Weber, I.T. 2005, Molecular basis for substrate recognition and drug resistance from 1.1 to 1.6 angstroms resolution crystal structures of HIV-1 protease mutants with substrate analogs. *FEBS J*. **272**: 5265–5277.

Tözsér, J., Oroszlan, S. 2003, Proteolytic events of HIV-1 replication as targets for therapeutic intervention. *Curr Pharm Des*. **9**: 1803–1815.

Tözsér, J., Weber, I.T. 2007, The protease of human T-cell leukemia virus type-1 is a potential therapeutic target. *Curr Pharm Des*. **13**: 1285–1294.

Tyndall, J.D., Pattenden, L.K., Reid, R.C., Hu, S.H., Alewood, D., Alewood, P.F., Walsh, T., Fairlie, D.P., Martin, J.L. 2008, Crystal structures of highly constrained substrate and hydrolysis products bound to HIV-1 protease. Implications for the catalytic mechanism. *Biochemistry*. **47**: 3736–3744.

van Maarseveen, N.M., Wensing, A.M., de Jong, D., Taconis, M., Borleffs, J.C., Boucher, C.A., Nijhuis, M. 2007, Persistence of HIV-1 variants with multiple protease inhibitor (PI)-resistance mutations in the absence of PI therapy can be explained by compensatory fixation. *J Infect Dis*. **195**: 399–409.

Wang, Y.-F., Tie, Y., Boross, P.I., Tozser, J., Ghosh, A.K., Harrison, R.W., Weber, I.T. 2007, Potent antiviral compound shows similar inhibition and structural interactions with drug resistant mutants and wild type HIV-1 protease. *J Med Chem*. 50: 4509–4515.

Watkins, T., Resch, W., Irlbeck, D., Swanstrom, R. 2003, Selection of high-level resistance to human immunodeficiency virus type 1 protease inhibitors. *Antimicrob Agents Chemother*. **47**: 759–769.

Weber, I.T. 1990, Comparison of the crystal structures and intersubunit interactions of human immunodeficiency and Rous sarcoma virus proteases. *J Biol Chem*. **265**: 10492–10496.

Weber, I.T., Kovalevsky, A.Y., Harrison, R.W. 2007, Structures of HIV protease guide inhibitor design to overcome drug resistance. In Frontiers in Drug Design and Discovery (Caldwell, G., Atta-ur-Rahman, Player, M., Choudhary, M., eds.), Vol. 3, pp. 45–62, Bentham Science Publishers, U.A.E.

Wiegers, K., Rutter, G., Kottler, H., Tessmer, U., Hohenberg, H., Krausslich, H.G. 1998, Sequential steps in human immunodeficiency virus particle maturation revealed by alterations of individual Gag polyprotein cleavage sites. *J Virol*. **72**: 2846–2854.

Wilson, W., Braddock, M., Adams, S.E., Rathjen, P.D., Kingsman, S.M., Kingsman, A.J. 1988, HIV expression strategies: ribosomal frameshifting is directed by a short sequence in both mammalian and yeast systems. *Cell*. **55**: 1159–1169.

Wlodawer, A., Gustchina, A. 2000, Structural and biochemical studies of retroviral proteases. *Biochim Biophys Acta*. **1477**: 16–34.

Wlodawer, A., Vondrasek, J. 1998, Inhibitors of HIV-1 protease: a major success of structure-assisted drug design. *Annu Rev Biophys Biomol Struct*. **27**: 249–284.

Zennou, V., Mammano, F., Paulous, S., Mathez, D., Clavel, F. 1998, Loss of viral fitness associated with multiple Gag and Gag-Pol processing defects in human immunodeficiency virus type 1 variants selected for resistance to protease inhibitors in vivo. *J Virol*. **72**: 3300–3306.

Chapter 3
Hepatitis C Virus

Philip Tedbury and Mark Harris

Abstract Hepatitis C virus is an important human pathogen infecting an estimated 170 million individuals. It is an enveloped virus with a positive sense RNA genome encoding a single polyprotein. This is cleaved by both cellular and viral proteases to yield ten mature polypeptides – both structural and non-structural (NS). The virus encodes two distinct proteases: firstly the NS2/3 autoprotease consisting of the C-terminus of NS2 and the N-terminus of NS3 – this cleaves itself at the junction between the two proteins. The N-terminus of NS3 then constitutes a distinct protease that cleaves at the junctions between the remaining non-structural proteins that form the C-terminal half of the polyprotein. In this article we describe our current understanding of the biochemistry and structural biology of NS2/3 and NS3 cleavage. We also discuss their roles in the viral life cycle, and highlight the current and future development of antiviral therapy targeted to these protease activities.

Keywords hepatitis C virus • NS2/3 autoprotease • NS3 serine protease • polyprotein processing • antiviral therapy

3.1 Introduction

3.1.1 Pathology of Hepatitis C Virus

Hepatitis C virus (HCV) was identified in 1989 by screening a cDNA expression library produced from the serum of chimpanzees infected with non-A non-B hepatitis with patient anti-sera (Choo et al., 1989). This discovery also allowed the development of serological tests for diagnosis of the infection and it is currently estimated that 2.2% of the global population are infected with HCV (Kuo et al., 1989; The Global Burden of Hepatitis C Working Group, 2004). HCV is transmitted

P. Tedbury and M. Harris (✉)
Institute of Molecular and Cellular Biology, Faculty of Biological Sciences, University of Leeds, Leeds. LS2 9JT, UK
e-mail: M.Harris@leeds.ac.uk

U. Lendeckel and N.M. Hooper (eds.), *Viral Proteases and Antiviral Protease Inhibitor Therapy*, Proteases in Biology and Disease 8,

parenterally and is spread most efficiently through intravenous drug use and contaminated blood products, although transmission via organ transplantation, tattooing and the use of un-sterilised needles in mass vaccination programs also plays a part. In developed countries intravenous drug use has taken over from transfusion as the main route of transmission; in a UK study HCV seropositivity was found in 7% of intravenous drug users who had been injecting for less than 3 years and 62% of those who had been injecting for 15 years or more (Hope et al., 2001).

Clearance of HCV infection requires the combined actions of the innate and adaptive immune systems and is associated with a rapid and prolonged response against a broad range of viral antigens by T-helper (Th) 1 $CD4^+$ T-cells (Day et al., 2002; Thimme et al., 2002). This immune response is critical for the recruitment and activation of $CD8^+$ or cytotoxic T-cells (Grakoui et al., 2003), which actively kill infected hepatocytes and produce antiviral cytokines such as IFN γ. HCV in turn escapes the immune response via a range of mechanisms including: escape mutants generated by error-prone replication; blockade of IFN signalling within the cell; and direct interference with T-cell and dendritic cell function (Foy et al., 2005; Hoofnagle, 2002; Racanelli and Manigold, 2007; Wedemeyer et al., 2002).

The ability of HCV to evade the immune response results in the frequent progression to chronic disease, defined as the persistence of viral RNA for at least 6 months, after which time spontaneous clearance is unusual (Yokosuka et al., 1999). The early symptoms are similar for acute and chronic infections and are usually sub-clinical. Although a variety of factors, including milder symptoms, being male, the lack of a vigorous T-cell response and increased age, are often associated with the development of chronic infection, no specific indicators have been identified to reliably predict chronic disease progression (Chen and Morgan, 2006). Once chronic infection is established there are few if any symptoms until the liver begins to fail, usually many years after the initial infection (Hoofnagle, 1997). Death from chronic HCV infection is a result of decompensated cirrhosis, where the level of liver function remaining is insufficient to support life, or the development of hepatocellular carcinoma (Chen and Morgan, 2006).

Therapeutic options for the treatment of HCV are limited, with a combination therapy of IFN and ribavirin (a guanosine analogue) forming the main part of most regimens. Successful treatment is most likely if begun before the onset of decompensated liver disease and in the absence of complicating factors such as HIV co-infection, immune suppressive diseases and alcohol abuse (Weigand et al., 2007). The genotype of HCV also has a large influence on the outcome of treatment; sustained virological responses (where the virus is reduced below detectable levels and remains undetectable following the cessation of treatment) of 55% are reported for genotype 1 infections, while 80% or higher is frequently seen with genotype 2 and 3 infections (Hadziyannis et al., 2004). In addition to the risk of a failure to respond to treatment, side effects are common with IFN and include influenza-like symptoms such as fatigue and muscle pain, in addition to depression and sleep disturbance. The most common side effect of ribavirin therapy is haemolysis leading to anaemia, consequently patients are monitored monthly for signs of toxicity;

nevertheless, non-compliance is a significant problem (Hughes and Shafran, 2006; Weigand et al., 2007). For these reasons new anti-HCV agents are highly desirable, the principle novel targets being the RNA-dependent RNA polymerase (RdRp) and the protease (Carroll and Olsen, 2006; Reesink et al., 2006).

3.1.2 Molecular Biology of HCV

On the basis of genome sequence and organisational similarity HCV is classified as a member of the *Flaviviridae* family and is the prototypic member of the genus *Hepaciviruses* which also encompasses the GB viruses (isolated from the serum of a surgeon George Barker) (Karayiannis and McGarvey, 1995). It therefore shares common features with *Flaviviruses* such as the Dengue Fever and Yellow Fever viruses, and *Pestiviruses* such as Bovine Viral Diarrhoea Virus (BVDV) (Major et al., 2001). It can be further separated into six major genotypes, and several subtypes (indicated by letter), with distinct clinical characteristics (Simmonds, 2004; Simmonds et al., 2005). For example, genotype 1 responds poorly to IFN therapy relative to genotypes 2 and 3, while genotype 3 infections are particularly associated with the development of hepatic steatosis (Hadziyannis et al., 2004; Rubbia-Brandt et al., 2000).

HCV has a positive-sense single-stranded RNA genome of approximately 9,600 nucleotides, flanked by 5′ and 3′ un-translated regions (UTRs) (Choo et al., 1991) (Fig. 3.1). Replication of the HCV genome takes place via the production of a complementary negative-strand which serves as a template for the production of

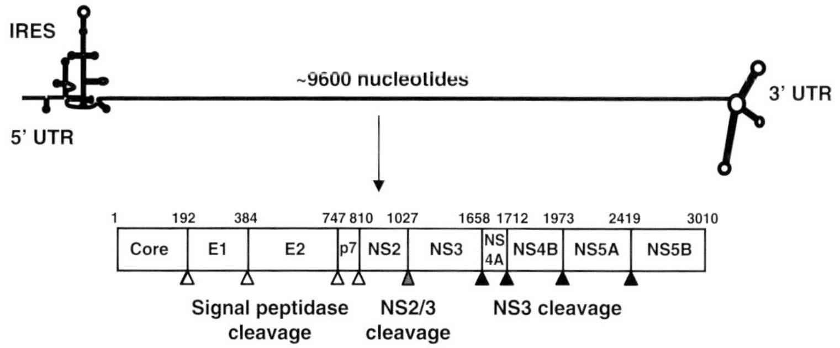

Fig. 3.1 The HCV genome and polyprotein. The genome is translated as a single polyprotein, the structural proteins located at the N-terminus and the non-structural proteins at the C-terminus. Distances along the polyprotein are shown in amino acid residues. UTR, untranslated region; IRES, internal ribosomal entry site; E, envelope glycoprotein; NS, non-structural protein. Processing of the polyprotein is by signal peptidase in the structural region and by the virus encoded protease NS3 in the non-structural region. The NS2–NS3 junction is cleaved by the NS2/3 autoprotease. Δ, signal peptidase cleavage site; ▲, NS2/3 autoprotease cleavage site; ▲, NS3 protease cleavage site

new positive-strand genomes (Lindenbach and Rice, 2005; Moradpour et al., 2007). The viral proteins NS3, NS4A, NS4B, NS5A and NS5B (RdRp) are required for the formation of the replication complex on modified cellular membranes, characterised by their resistance to disruption by non-ionic detergents (Shi et al., 2003). Although NS5B is capable of copying non-HCV templates, the replication complex is able to specifically identify and replicate the HCV genome via genotype specific recognition signals in the 5′ and 3′ UTRs which interact with NS3, NS5A and NS5B (Murayama et al., 2007). Prior to 2005, research into HCV replication was hindered by the inability to grow the virus in culture. Instead a great deal of work was performed using subgenomic replicons which incorporated the proteins NS3–NS5B, the 3′ and 5′ UTRs and a reporter gene such as neomycin phosphotransferase or firefly luciferase (Krieger et al., 2001; Lohmann et al., 1999). In 2005 three groups were able to demonstrate the complete virus life-cyle, replication, particle production and re-infection of cultured cells and chimpanzees (Lindenbach et al., 2005; Wakita et al., 2005; Zhong et al., 2005). This was possible due to the discovery of the isolate Japanese Fulminant Hepatitis-1 (JFH-1), acquired from a patient with an acute infection and able to replicate to much high levels than previously available isolates (Kato et al., 2003).

The genome contains a single open reading frame (ORF) that encodes a polyprotein of approximately 3,000 amino acids, the precise size depending on genotype. The structural proteins form the virus particle and comprise the capsid protein, Core, and the envelope glycoproteins, E1 and E2. A viroporin, p7, is also typically classified as a structural protein, although its presence in the virion has yet to be demonstrated. The structural proteins are located at the amino (N)-terminus of the polyprotein and are released by signal peptidase cleavage. The non-structural proteins are principally required for the replication of the virus and are located at the carboxyl (C)-terminus of the polyprotein. They have a wide range of functions in addition to replication of the HCV genome: NS2 forms part of the NS2/3 autoprotease; NS3 is the viral protease responsible for the cleavage of the downstream proteins and also possesses an RNA helicase domain; NS4A is a co-factor for NS3 protease function; NS4B is necessary and sufficient for formation of the membranous web upon which HCV replication occurs; NS5A has numerous functions in cell signalling; and NS5B is an RNA dependent RNA polymerase (Moradpour et al., 2007).

The proteases required in the HCV life-cycle fall into three distinct categories which can be discussed in turn: the host cell proteases which are required for cellular functions in addition to their roles in cleaving the structural region of the HCV polyprotein; the NS3 serine protease which cleaves the non-structural region of the polyprotein and some cellular targets; and the NS2/3 autoprotease.

3.2 Host Cell Proteases

The structural proteins are released from the polyprotein by signal peptidase cleavage. The core protein is further processed by signal peptide peptidase cleavage, from an immature 23 kDa peptide to the 21 kDa mature form (McLauchlan et al.,

2002; Yasui et al., 1998). A third, 16 kDa form has been observed but seems to be unique to the prototype, HCV-1 strain of the virus and even then is a major product only in the absence of the E1 coding sequence (Bukh et al., 1994; Lo et al., 1994).

Core is localised predominantly to the cytoplasmic face of the endoplasmic reticulum (ER) membrane and to lipid droplets (Moradpour et al., 1996; Rouille et al., 2006). The lipid droplet localisation is dependent on a hydrophobic region of domain 2, which is thought to insert into the ER membrane (Boulant et al., 2006; Hope and McLauchlan, 2000). The full length core is also anchored to the ER by the signal peptide preceding E1, which remains attached to core following separation from E1. Further cleavage of this region by signal peptide peptidase results in formation of the 21 kDa species and allows localisation to lipid droplets (McLauchlan et al., 2002). Blocking this processing step either pharmacologically or by mutation of the target sequence prevents both localisation to lipid droplets and virus particle formation. (Boulant et al., 2007; McLauchlan et al., 2002). Another study indicated that the maturation step may take place after budding, as the 23 kDa species was able to form more stable particles than the 21 kDa, suggesting that lipid droplet association is not required for particle formation, but that signal peptide peptidase may be incorporated into the particle and that maturation of core is needed to promote capsid disassembly during cell entry; however, these findings have not been verified using infectious virus (Vauloup-Fellous et al., 2006).

3.3 NS3 Protein

3.3.1 A Multifunctional Protein

The N-terminal third of NS3 (amino acids 1027–1206) possesses a chymotrypsin-like serine protease activity that is required for processing the non-structural region of the HCV polyprotein (Failla et al., 1994; Grakoui et al., 1993a; Tomei et al., 1993). The C-terminal two thirds contain both a helicase and an NTPase activity, and can be classified as part of the DEXH sub-family of the DEAD-box helicases (Kim et al., 1995). The protease and helicase activities display no dependence on one another for activity (Gallinari et al., 1998).

NS3 is the only HCV protein not to interact directly with membranes, it is nevertheless invariably found in association with membranes due to a non-covalent interaction with NS4A which serves as a membrane anchor and cofactor to the protease activity (Failla et al., 1994; Kim et al., 1999). In hepatocytes harbouring sub-genomic replicons, NS3 displays a perinuclear localisation and co-localises with the other non-structural proteins, presumably in replication complexes (Mottola et al., 2002).

The helicase activity is less well characterised than the protease, although it is known to unwind double-stranded RNA and DNA-RNA heteroduplexes, using ATP as an energy source and by a divalent metal ion dependent mechanism (Gwack et al., 1997; Suzich et al., 1993). NS3 helicase activity is not enhanced by NS4A binding, in fact there is evidence to suggest that the binding of NS4A induces a conformation of NS3 which in turn favours protease activity at the expense of helicase activity (Gallinari et al., 1999). The protease and helicase domains do demonstrate a level of interdependence, however, as NS3 protease requires the helicase domain for optimal function *in vitro* (Beran and Pyle, 2008).

3.3.2 NS3 Protease Mechanism

The crystal structure of the NS3 protease domain has been solved both in the presence and absence of NS4A (Kim et al., 1996; Love et al., 1996). These structures revealed three critical sites within the NS3 protease, namely the active site, a zinc-binding site and the NS4A binding site (Fig. 3.2). The active site corresponds

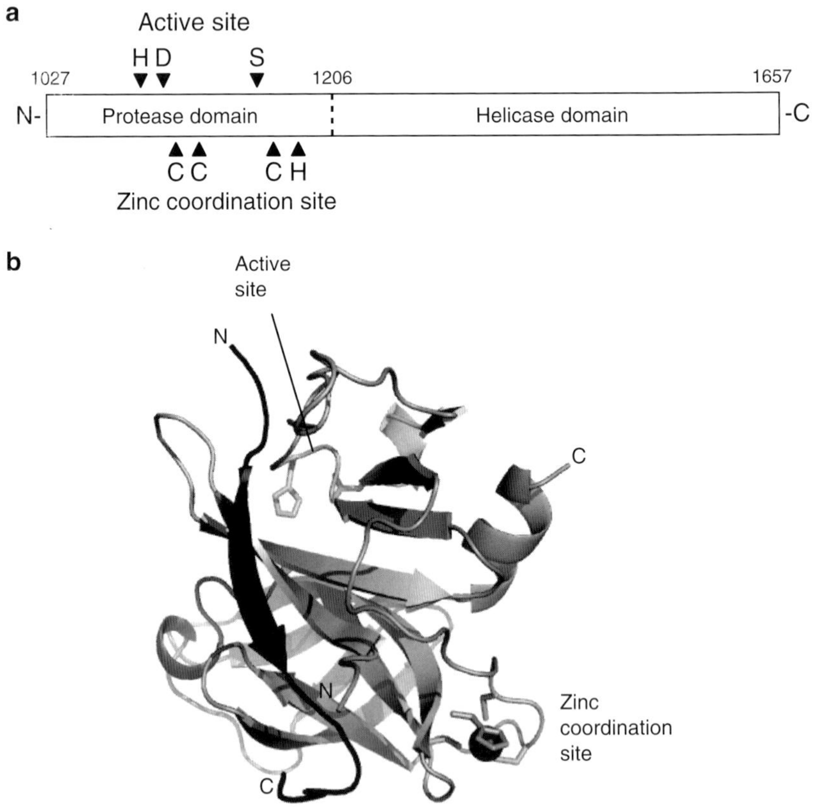

Fig. 3.2 The NS3 protein. (**a**) Schematic of NS3. The protease and helicase domains are shown with the residues of the active site (H1083, D1107 and S1165 – numbered according to the Con1 polyprotein) and zinc coordinating site (C1121, C1123, C1171 and H1175). (**b**) Structure of the NS3 protease domain. The polypeptide comprising the NS3 protease domain is shown in gray with the N and C termini indicated. The active site and zinc coordinating residues are shown with side chains and the zinc ion in black. The NS4A protein is shown in black with the N and C termini indicated. Structure from Kim et al. (1996). Image generated using PyMOL version 0.99 (DeLano Scientific LCC)

to histidine 1083, aspartic acid 1107 and serine 1165 (numbering according to the genotype 1b Con1 polyprotein). The structure solved in the absence of NS4A revealed an atypical conformation of the active site residues, wherein the imidazole ring of H1083 is too far from S1165 to deprotonate the nucleophilic hydroxyl group and the side chain of Asp1107 is oriented away from H1083 and is therefore unable to effect charge stabilisation following acquisition of the proton by histidine (Love et al., 1996). NS4A binds to NS3, fitting between two β-strands of the protease domain, by virtue of hydrophobic interactions, and promotes a more tightly packed protease domain (Kim et al., 1996). This structural change requires only a short peptide from the centre of NS4A (residues 21–32) and both stabilises NS3 and enhances its protease activity (Bartenschlager et al., 1995; Gallinari et al., 1999; Kim et al., 1996; Lin et al., 1995). Enhancement of activity is believed to be mediated by changes at the level of tertiary structure as the active site residues adopt a classical catalytic triad configuration (Bianchi et al., 1997; Steinkuhler et al., 1996). There is evidence that junctions within the non-structural region are processed at different rates and exhibit different sensitivity to the presence of NS4A. NS5A–NS5B and NS4A–NS4B are cleaved rapidly and can be cleaved in the absence of NS4A. In addition to the general increase in NS3 protease activity, NS4A is particularly required for processing of the NS3–NS4A and NS4B–NS5A junctions (Bartenschlager et al., 1994; Failla et al., 1994; Lin et al., 1994).

The zinc ion is coordinated by the three cysteine residues (1121, 1123 and 1171) and, via a water molecule, histidine 1175. This zinc has been shown to be required for a properly folded and active protease (De Francesco et al., 1998; Stempniak et al., 1997). The zinc ion is located at least 20 Å from the catalytic serine and studies on the rhinovirus 2A cysteine protease, which possesses a similar zinc binding motif to NS3, showed zinc to be required for structural integrity rather than catalysis (Sommergruber et al., 1994; Yu and Lloyd, 1992); the role of zinc in NS3 is consequently thought to be entirely structural.

3.3.3 NS3 Protease and Innate Immunity

In addition to its critical role of processing the HCV polyprotein, NS3 has been reported to have a variety of functions regulating the biology of the host cell, including: effects on signalling pathways due to repression of phosphorylation by protein kinase A (Borowski et al., 1999; Aoubala et al., 2001); induction of apoptosis via caspase 8 (Prikhod'ko et al., 2004); stimulation of growth via the cJun N-terminal kinase (JNK) pathway (Hassan et al., 2005); and transformation of NIH 3T3 cells and rat fibroblasts (Sakamuro et al., 1995; Zemel et al., 2001).

A further, recently identified and increasingly well characterised, function of NS3 is a role in suppressing the host IFN response. Type 1 IFN (α and β) are cytokines generated in response to viral infection and serve to induce an "antiviral state" in neighbouring cells. Type 1 IFN production is typically stimulated by the detection of viral RNA by one of two principle cellular mechanisms: the Toll-like receptor (TLR) 3 and retinoic acid-induced gene (RIG)-I pathways, which have been shown to recognise and respond to HCV RNA (Li et al., 2005a; Saito et al., 2008). NS3 is able to cleave adaptor molecules from both of these signalling pathways, Toll-IL-1 receptor domain containing adaptor-inducing IFN-β (TRIF) and caspase recruitment domain adaptor inducing IFN-β (CARDIF, also known as IPS-1, MAVS and VISA) respectively, due to similarities to the NS3 consensus cleavage sequence, loosely defined as (D/E)xxxx(C/T)(S/A) with cysteine or threonine at the P1 position. Consequently the induction of IFN-β production is severely impaired, due to a blockade in the activation of a variety of transcription factors involved in the innate immune response (Breiman et al., 2005; Foy et al., 2005; Li et al., 2005b). The importance of these pathways in suppressing viral replication is underlined by the ability of the NS3 protease of the related Hepacivirus, GBV-B, to cleave CARDIF and the enhanced HCV replication in Huh 7.5 cells, which have a mutation in the CARD domain of RIG-I, which abrogates the IFN-β and early IFN stimulated gene (ISG) response, rendering them far more permissive to HCV replication than the parental Huh7 cell line (Chen et al., 2007; Sumpter et al., 2005).

3.3.4 NS3 Protease as a Drug Target

Following the success of drugs targeting the HIV protease, NS3 protease is one of the most promising HCV specific targets. The development of potent HIV protease inhibitors was made possible by the use of *in vitro* protease assays and high-throughput screening; similarly the HCV NS3 protease can be assayed using an in vitro system based on purified NS3 and short synthetic substrate peptides (Kakiuchi et al., 1999). These peptides are labelled with a fluorophore and quencher at either terminus such that when un-cleaved they do not fluoresce because energy emitted by the fluorophore is absorbed by the quencher in a manner akin to fluorescence resonance energy transfer (FRET). Following cleavage, an increase in fluorescence proportional to NS3 protease activity can be measured.

Unlike many proteases, NS3 has a shallow substrate binding site which does not obviously lend itself to the design of high affinity competitive inhibitors (Kim et al., 1996). Several compounds have nevertheless been developed and have progressed into clinical trials. One such compound, BILN 2061, showed potent anti-NS3 activity *in vitro* and in the replicon system and represented the first specific anti-HCV drug to reach patient trials. Its excellent efficacy against genotype 1 NS3 *in vitro* was mirrored in patients, bringing about a 2–3 $\log_{10}$ reduction in serum levels of viral RNA within 24 h. This efficacy was combined with low cellular toxicity and no apparent adverse responses in trials with healthy human volunteers (Hinrichsen et al., 2004; Lamarre et al., 2003). The success of the BILN 2061 in patients was arguably even greater than its *in vitro* activity suggested and hinted at a possible second and synergistic antiviral effect. As discussed above, NS3 is plays a significant role in the suppression of the innate antiviral response by cleaving two proteins, TRIF and CARDIF, which are required for initiation of the IFN response to dsRNA. As a consequence, inhibition of NS3 protease not only exerts a direct antiviral response by blocking polyprotein processing but also exerts an indirect antiviral response by relieving the blockade of the IFN response.

Although humans trials with BILN 2061 were abandoned on safety grounds, after cardiotoxicity was seen in Rhesus monkeys subject to high doses (Hinrichsen et al., 2004; Reiser et al., 2005), it had demonstrated the potential impact of NS3 proteases in HCV therapy. Other compounds have followed, including SCH503034 (Sarrazin et al., 2007), VX-950 (Reesink et al., 2006) and ITMN-191 (Seiwert et al., 2006); all have shown efficacy in patients and of which VX-950 is the most advanced through clinical trials. These clinical trials have also served to highlight one of the greatest difficulties in developing therapies for viruses such as HCV. The highly error prone nature of the viral polymerase generates a quasispecies in the patient (Ogata et al., 1991). This quasispecies contains an immense potential for resistant mutants to any drug used. All of the protease inhibitors trialled thus far have lead to the rapid appearance of resistance and viral rebound. To counter this resistance it is necessary to develop numerous effective inhibitors, ideally with a variety of modes of action. The inhibitors described so far all act on the NS3 active site, and although they possess differing resistance profiles, mutants have been described with significant resistance to multiple compounds. At present there is only one inhibitor with a truly different mode of action, ACH-806, which inhibits HCV by binding to NS4A, the NS3 protease cofactor. This is believed to result in the inefficient processing of the HCV polyprotein and failure to form functional replication complexes (Yang et al., 2008). Although resistant mutants to this drug have been identified, exclusively in NS3, none of these mutations exhibit cross-resistance with other HCV inhibitors (Yang et al., 2008).

3.4 NS2/3 Autoprotease

3.4.1 Mechanism of NS2/3 Processing

Studies into polyprotein processing in HCV revealed that cleavage of the structural proteins was by signal peptidase and of the non-structural proteins by an NS3 encoded serine protease; however, mutations to the NS3 active site had no impact on processing at the NS2–NS3 junction (Grakoui et al., 1993a). Instead, work performed in mammalian and bacterial cells or using *in vitro* translation systems, demonstrated that processing at this site was dependent on a second protease activity (Grakoui et al., 1993b; Hijikata et al., 1993). Truncation analysis determined that this second protease activity required both NS2 (Fig. 3.3) and the protease domain of NS3 and was therefore termed the NS2/3 autoprotease (Fig. 3.4). Further truncations revealed that the predicted hydrophobic N-terminus of NS2 was dispensable for enzymatic activity, while deletion of even the 9aa C-terminal alpha-helix from the NS3 protease domain was sufficient to completely prevent NS2/3 processing

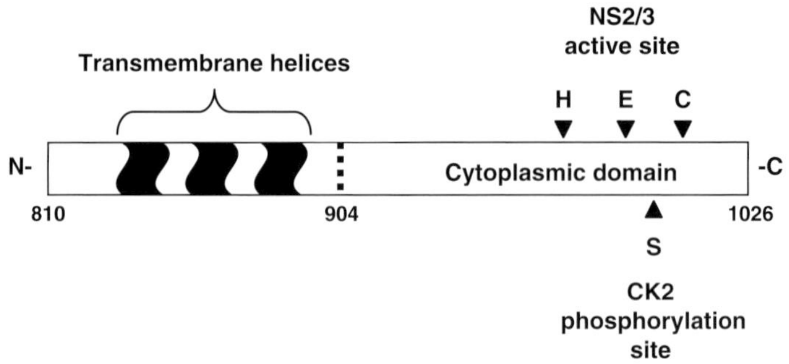

Fig. 3.3 Schematic of the NS2 protein. The N-terminus of the protein is processed by signal peptidase and resides in the ER lumen, while the C-terminus is processed by the NS2/3 autoprotease and resides in the cytoplasm. The cytoplasmic domain contains the NS2/3 autoprotease active site (H952, E972 and C993) and a CK2 phosphorylation site (S977)

(Pallaoro et al., 2001; Tedbury and Harris, 2007; Thibeault et al., 2001). Within this minimal active region (Fig. 3.4) the active site was further delineated as H952, E972 and C993 (Grakoui et al., 1993b; Hijikata et al., 1993). Although not part of the active site, the residue C922 located towards the N-terminus of the NS2 cytoplasmic domain is also required for NS2/3 activity both *in vitro* and in replicons, as mutation to alanine or serine in either system significantly abrogates activity (Tedbury and Harris, 2007). However, at present it has not been possible to explain why it is required for activity.

While the structure of the NS2/3 autoprotease as yet defies solution, the crystal structure is available for the NS2 cytoplasmic domain post-cleavage allowing NS2/3 to be classified as a novel cysteine protease (Lorenz et al., 2006). Perhaps the most surprising feature of the structure was the dimeric nature, not only of the protein, but of the active site. The residues H952 and E972 of the first monomer combine with C993 in the second to create an active site which cleaves the second monomer, and vice versa (Fig. 3.4b). This mechanism, although difficult to visualise in the context of two separate polyproteins having to undergo cotranslational dimerisation to allow processing, was confirmed by experiments combining separate proteases with either the alanine substitution of H952A or C993A.

These constructs were able to form heterodimers containing one functional active site and one non-functional active site with both substituted residues. The C-terminal leucine of NS2 remains in the active site following cleavage, presumably serving to inactivate the protease and prevent non-specific proteolytic activity. This structure partially explains the mechanism of NS2/3 processing and the bi-molecular cleavage described in previous papers, and both the detection of dimers and the measurement of an optimum active concentration, which are otherwise difficult to explain in an intra-molecular cleavage model (Grakoui et al., 1993b; Pallaoro et al., 2001; Reed et al., 1995). In addition, the reduction in folding efficiency and dimer formation observed by Pallaoro et al. when using constructs containing the mutation C993A, may explain the low efficiency of *trans*-complementation.

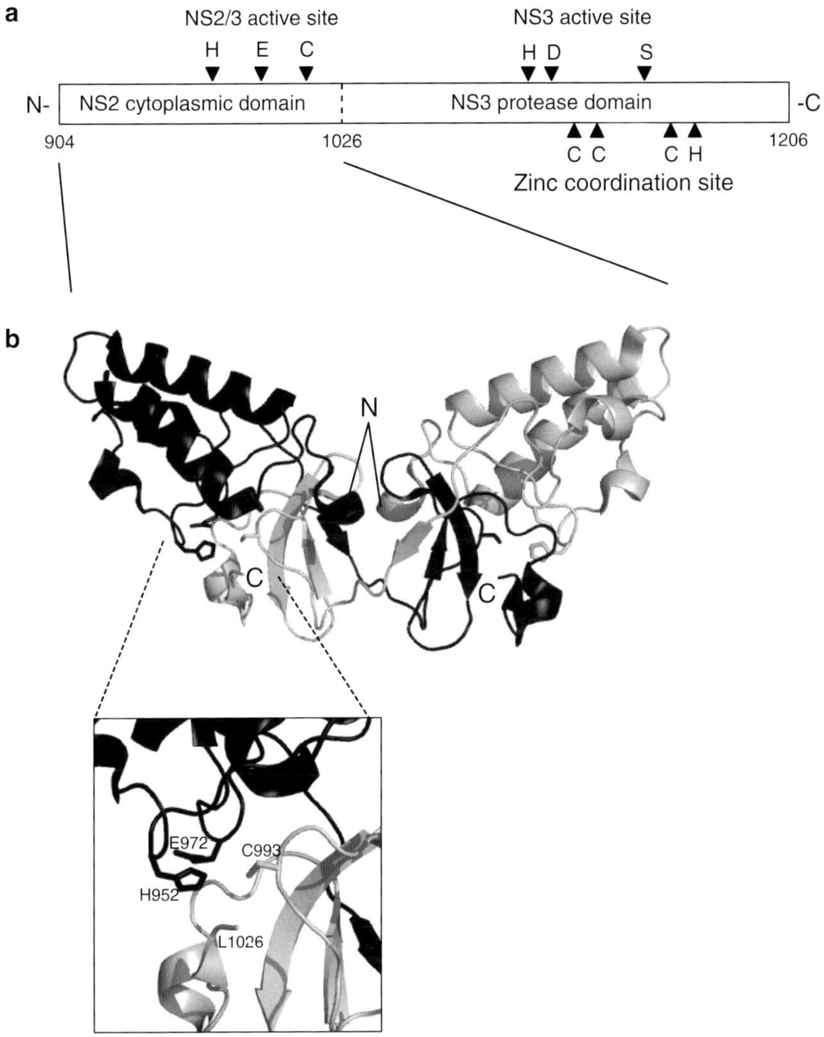

Fig. 3.4 The NS2/3 autoprotease. (**a**) Schematic of the minimal requirement for NS2/3 autocleavage. The cytoplasmic domain of NS2 (from residue 904) together with the protease domain of NS3 (to residue 1206) can be expressed and purified from *E. coli* and contains all the requirements for autocleavage in vitro. Although the entire NS3 protease domain is required for autocleavage, NS3 protease activity is not. (**b**) Structure of the catalytic domain. The NS2 cytoplasmic domain dimer is shown with Chain A shown in gray and Chain B in black, and N and C termini indicated. The active site residues H952, E972 and C993 are shown with side chains. The active site of chain A is shown enlarged as a composite of H952 and E972 from chain B chain and C933 and L1026 from chain A. Structure from Lorenz et al. (2006). Image generated in Pymol version 0.99

Table 3.1 Characteristics of NS2/3 and NS3 ± NS4A

	NS3 + NS4A	NS3 – NS4A	NS2/3
Affinity for Zn	High	Low	Low
H1175 required for Zn coordination	Yes	No	No
H1175 required for activity	Yes	No	No

NS3 is not present in the crystal structure of the NS2/3 catalytic domain; it is, however, required for the autoproteolytic activity. Mutational analysis of the zinc coordinating site of NS3 demonstrated that NS3 and NS2/3 proteases share a dependence on the zinc ion coordinated by NS3. Furthermore, detailed characterisation of zinc binding and activity in the two proteases in the presence and absence of NS4A peptide revealed that NS2/3 was more closely comparable to NS3 in the absence of NS4A than in its presence (Table 3.1) (Tedbury and Harris, 2007). The binding of NS4A to NS3 has been shown to alter or stabilize the conformation of the protease domain; the two conformations are referred to as "open" (without 4A) and "closed" (with 4A) (Love et al., 1998). Zinc binding is also affected by the presence of NS4A, the affinity of NS3 for zinc being much increased by NS4A, apparently due to effects on the orientation of H1175 towards the zinc ion. It can be surmised that the differences observed in affinity for zinc are the consequence of the two conformations, where open has a low affinity for zinc and closed a high affinity. It can then be hypothesised that the open conformation adopted in the absence of NS4A and seemingly required by NS2/3 autoprotease uniquely presents the scissile bond in the manner necessary for cleavage. This hypothesis would also explain the inhibition of NS2/3 cleavage by high concentrations of NS4A, as the binding of NS4A will force the NS3 domain into the closed conformation, removing the scissile bond from the active site of NS2/3 (Darke et al., 1999; Thibeault et al., 2001).

Analysis of the NS2/3 cleavage site revealed that although highly conserved, mutations to the residues comprising the cleavage site have little impact on processing efficiency (Reed et al., 1995). The most important residues are those at positions P1 and P1′, leucine and alanine respectively; if these residues are replaced with proline or aspartic acid cleavage is abrogated. The requirement for small hydrophobic residues suggests that the inhibition by bulky or charged residues is due to disruption to folding of the cleavage site or entire protein. This idea was supported by investigations into *trans*-cleavage of NS2/3 precursors and the loss of NS3 protease activity from NS2/3 precursors bearing the mutation of the cleavage site to two proline residues (Grakoui et al., 1993b; Reed et al., 1995; Tedbury and Harris, 2007). Wildtype NS2/3 autoprotease can cleave a sequence bearing a mutation of either H952 or C993 to alanine; however, this is dependent on the substrate protein containing the complete sequence required for cleavage: truncations to the N- or C-termini that prevent autoprocessing also prevent *trans*-cleavage by the wildtype protease. These findings reiterate the importance of correct folding for the autocleavage and suggest that it is the loss of this folding that prevents autocleavage in the truncated proteins, even where the active site and cleavage site remain intact.

3.4.2 NS2/3 Processing in the Virus Life-Cycle

It remains unclear why the NS2–NS3 junction is processed in the manner it is in HCV. In Flaviviruses, which share a similar genome and polyprotein organisation to HCV, the NS2–NS3 junction is cleaved by NS3 and the flaviviral NS2B then plays a role analogous to HCV NS4A (Lindenbach and Rice, 2001). Pestiviruses possess an NS2–NS3 junction which can be cleaved by an NS2/3 autoprotease similar to that found in HCV. In BVDV temporal regulation of NS2/3 and NS3 levels by the abundance of a cellular chaperone, DNAJ, is involved in the regulation of cytopathic and non-cytopathic biotypes of the virus (Lackner et al., 2004; Rinck et al., 2001); while there is evidence suggesting a role for the chaperone HSP90 in HCV NS2/3 processing, there is little evidence of regulation, as processing seems to be invariably efficient and uncleaved NS2/3 is not observed in infected cells, either in acute or chronic infections (Waxman et al., 2001). As the closest relatives of HCV, the GB viruses, are little studied, the best indications as to the role of NS2/3 and NS3 in HCV are studies on that virus itself (Simons et al., 1995a, b).

Unique amongst the non-structural proteins of HCV, NS2 itself seems to have no role in genome replication. Indeed, when sub-genomic replicons were produced with and without NS2, those without replicated more efficiently (Lohmann et al., 1999). NS2/3 processing, however, is required; when NS2/3 processing was knocked-out in sub-genomic replicons, replication was completely abrogated, as was that of a similarly mutated virus, despite uncleaved NS2/3 precursors retaining both protease and helicase activity (Jones et al., 2007; Welbourn et al., 2005). To further analyse the role played by NS2 protease viruses were generated where Core–NS2 and NS3–NS5B were expressed as separate ORFs, with translation of the second ORF driven by an IRES (Jones et al., 2007). In this system it could be shown that mutations to the NS2 protease active site had no effect on replication or virion production, indicating in turn that the loss of replication associated with a loss of NS2/3 processing is most likely due to the effect on NS3, a hypothesis supported by the inhibition of the NS4A–NS3 interaction by uncleaved NS2 (Welbourn et al., 2005). Although NS2 protease has not been associated with any activity beyond the autoprotease and seems to function solely to process the NS2–NS3 junction, viruses lacking the NS2 protein or protease domain are able to replicate but unable to assemble new virions, suggesting a role in virus assembly (Jones et al., 2007).

The mechanism by which NS2 influences particle formation remains obscure. The efficient replication of HCV sub-genomic replicons in the absence of NS2 demonstrates that it is not an essential component of the replication complex. NS2 has been shown to interact with all the other non-structural proteins, however, and in light of its necessity for infectious particle formation may be hypothesised to play a mediating role between the structural proteins and the replication complexes (Dimitrova et al., 2003; Flajolet et al., 2000; Kiiver et al., 2006). Evidence for interactions between NS2 and the structural proteins includes p7-dependent re-direction of NS2 trafficking and localisation, from a diffuse to a punctate distribution, closely associated with replication complexes (Tedbury et al., manuscript in preparation) and a study using intergenotypic HCV chimeras fused at various sites in p7 and NS2; all produced infectious particles, but did so with the greatest efficiency when the fusion was located in NS2 such that the first transmembrane domain of NS2 was of the same genotype as the structural proteins (Pietschmann et al., 2006). A recent study demonstrated the importance of two further regions of NS2 in virus assembly – the transmembrane domains and the casein kinase 2 (CK2) target serine 977 (Franck et al., 2005; Jirasko et al., 2008). NS2 is believed to possess three transmembrane domains; deletion of either the first two or the second (to retain overall protein topology) blocked virus production, but had no impact on replication. Similarly mutation of S977 to alanine reduced virus production by tenfold, but did not affect replication. The specific role(s) of NS2 in assembly remains to be elucidated but could include interactions with other HCV proteins or the recruitment of cellular proteins or membranes.

3.4.3 Cellular Interactions of NS2

One of the first reports of host cell regulation by NS2 was that NS2 inhibited gene expression from viral promoters, including those from cytomegalovirus (CMV), SV-40 and hepatitis B virus (HBV), and from cellular promoters including TNF-α, ferrochelatase and NF-κB binding regions (Dumoulin et al., 2003). As some of the cellular promoters affected are involved in innate immunity and inflammation it is possible that NS2 is playing a role in regulating this aspect of cellular activity. A study of the effect of NS2 and NS3 on the IFN-β promoter demonstrated suppression of IFN-β expression by NS2, although unlike NS3/4A this was not due to cleavage of CARDIF nor was it a specific effect, as suppression of the control promoter was also seen (Kaukinen et al., 2006). Rather, these findings appeared to show that NS2 has a generally suppressive effect on host gene expression, which includes genes involved in the immune response. Inhibition of gene expression by NS2 can be mediated via stimulation of cAMP-dependent pathways, destabilising several cellular mRNAs (Kim et al., 2007). Inhibition of cyclin A resulting in cell cycle S-phase arrest has also been reported (Yang et al., 2006). Cyclin A mRNA contains a cAMP-response element and is regulated by JunB, the expression of which is regulated in turn by NF-κB (Andrecht et al., 2002; Lopez-Rovira et al., 2000). Owing to the diverse range of promoters apparently downregulated by NS2, it is difficult to discern which are genuinely critical to HCV replication.

At the present time, only one direct interaction has been described between NS2 and a cellular protein, that with cell death-inducing DNA fragmentation factor (DFF) 45-like effector (CIDE)-B (Erdtmann et al., 2003). CIDE-B induces apoptosis via mitochondrial release of cytochrome c and activation of caspase 3; NS2 interacts with CIDE-B in cells, down regulate its abundance and suppress cytochrome c release and apoptosis (Erdtmann et al., 2003). A subsequent report suggested a mechanism for this effect, demonstrating that NS2 is phosphorylated at S977 and subsequently subject to proteosomal degradation (Franck et al., 2005). When combined with an interaction between NS2 and CIDE-B this data suggests that CIDE-B may also be degraded by this route, inhibiting the induction of apoptosis in HCV infected cells.

3.4.4 NS2/3 as an Antiviral Target

Although NS2/3 has desirable features as a drug target, such as being required for viral replication and having low homology to any known human proteins, progress has been hindered by the lack of a high-throughput *in vitro* assay. Compounds have been identified which inhibit NS2/3 and they can be crudely separated into two categories, divalent metal ion chelators, such as EDTA and 1,10-phenanthroline, which inhibit by depriving the protease of the essential zinc ion, and agents such as iodoacetamide and N-ethylmaleimide, which chemically modify reactive amino acid side-chains (Pieroni et al., 1997; Thibeault et al., 2001). None of these are specific to NS2/3 autoprotease and offer little insight into potential therapies; in fact more specific inhibitors of cysteine or zinc-metalloproteases have not demonstrated activity against NS2/3 (Pieroni et al., 1997; Tedbury and Harris, 2007; Thibeault et al., 2001).

A high-throughput cell-based assay for NS2/3 has been described by (Whitney et al., 2002). It uses NS2/3 processing to remove a ubiquitin based destabilisation domain from β-lactamase (BLA) in a stable cell line. The assay thereby couples the readily measurable BLA activity to that of NS2/3 autoprotease. This method has the advantage of being performed in a cellular environment and in a multi-well plate format, and has been used to validate the anti-NS2/3 activity of HSP90 inhibitors, radicicol and geldinamycin (Waxman et al., 2001). However, it is a complex protocol which requires careful execution to produce reliable data and, because it is a cell based assay, compounds which are not membrane permeable will generate negative results even if they possess potent inhibitory activity against NS2/3 autoprotease, while compounds which inhibit translation or exhibit toxicity to cells will appear as hits. These problems could be alleviated with an *in vitro* screen, allowing the identification of a broad range of active compounds.

Methods have been developed to express and purify recombinant NS2/3 from *Escherichia coli*. Expression as inclusion bodies followed by solubilisation and purification under denaturing conditions avoids premature autocleavage, and activity can then be induced by dilution into an appropriate refolding buffer (Pallaoro et al., 2001; Thibeault et al., 2001). In order to assess the efficacy of cleavage it is necessary to separate the products by SDS-PAGE or HPLC, laborious steps which severely impair the throughput of the assay. To develop NS2/3 further as a drug target, these aspects of the assay would need to be replaced with higher throughput techniques. A fluorescent substrate based approach such as that used for NS3 would be ideal, however the autoproteolytic nature of NS2/3 protease may block access of a synthetic peptide to the active site, while it is anticipated that the proteolytic activity of NS2 post cleavage form is inhibited by the maintenance of its C-terminus in the active site (P. Tedbury, P. Thommes, and M. Harris, unpublished observations) (Lorenz et al., 2006). Although a high-throughput *in vitro* assay has not yet been described, one could be developed if an automated method could be developed for separating un-cleaved NS2/3 from its cleavage products.

3.5 Conclusions

HCV produces all of its proteins as a single polyprotein, which is then processed by host and virally encoded proteases to generate the ten mature protein products. The host signal peptidase and signal peptide peptidase are required the processing of the HCV structural proteins, but are unlikely drug targets owing to their critical roles in host cell biology.

NS3 and NS2/3 proteases are required for the processing of the HCV non-structural proteins and mutagenesis studies have demonstrated the requirement of the virus for their function in order to replicate. The folding of both proteases is dependent upon a zinc ion coordinated in the protease domain of NS3. In the case of NS3, the availability of detailed structural information and high-throughput screening approaches has led to the development of highly effective inhibitors suitable for progression into clinical trials. The clinical efficacy of these inhibitors is enhanced by inhibiting not only the essential role NS3 plays in polyprotein processing, but also its suppression of the host immune response.

Structural studies of the NS2/3 precursor, prior to cleavage, are hindered by the nature of the protein: when correctly folded it rapidly cleaves. Due to the dimeric structure of the catalytic domain, mutagenesis of the active or cleavage sites appears to cause a loss of efficient folding in addition to a loss of activity. Although it is possible to crystallise proteases in the presence of an inhibitor to prevent any autoproteolytic activity, the only available inhibitors of NS2/3 cleavage appear to function by disrupting the correct structure (T Foster, personal communication). A number of possibilities nevertheless exist for the development of NS2/3 as a drug target. High-throughput screening can be applied to find lead compounds, which could be refined using the structure of the catalytic domain of NS2. Secondly, although the structure of NS2/3 is not known, it is known that NS2/3 processing is inhibited when NS3 is bound to NS4A and adopts its "closed" conformation. In principle compounds could be developed to bind in place of NS4A and promote the structural alterations associated with NS2/3 inhibition while simultaneously blocking the NS3/4A interaction required for formation of the replication complex. A final possibility would be zinc ejection compounds; these could be active against both proteases, but as the zinc ion appears to be relatively exposed in NS2/3, compared to NS3, it would be more susceptible to this approach.

Current HCV therapy lacks specifically targeted anti-viral drugs, has limited efficacy and frequently causes unpleasant side effects. At present the major targets for drug development are the NS3 protease and NS5B polymerase, partly due to their vital functions within the virus but also due to the availability of high-throughput assays. NS2/3 has desirable features as a drug target, in spite of the difficulties it remains an extremely worthwhile goal: the experience with HIV clearly demonstrates that when dealing with RNA viruses and the high rate of mutation they exhibit, a wide range of therapeutic options are required and ideally used in combination.

Acknowledgements Research in our laboratory is funded by the Wellcome Trust (grant number 082812).

References

Andrecht, S., Kolbus, A., Hartenstein, B., Angel, P., Schorpp-Kistner, M. 2002, Cell cycle promoting activity of JunB through cyclin A activation. *J Biol Chem* **277:** 35961–35968.

Aoubala, M., Holt, J., Clegg, R.A., Rowlands, D.J., Harris, M. 2001, The inhibition of cAMP-dependent protein kinase by full length hepatitis C virus NS3/4A complex is due to ATP hydrolysis. *J Gen Virol* **82**: 1637–1646.

Bartenschlager, R., Ahlborn-Laake, L., Mous, J., Jacobsen, H. 1994, Kinetic and structural analyses of hepatitis C virus polyprotein processing. *J Virol* **68:** 5045–5055.

Bartenschlager, R., Lohmann, V., Wilkinson, T., Koch, J.O. 1995, Complex formation between the NS3 serine-type proteinase of the hepatitis C virus and NS4A and its importance for polyprotein maturation. *J Virol* **69:** 7519–7528.

Beran, R.K., Pyle, A.M. 2008, Hepatitis C viral NS3-4A protease activity is enhanced by the NS3 helicase. *J Biol Chem* **283:** 29929–29937.

Bianchi, E., Urbani, A., Biasiol, G., Brunetti, M., Pessi, A., De Francesco, R., Steinkuhler, C. 1997, Complex formation between the hepatitis C virus serine protease and a synthetic NS4A cofactor peptide. *Biochemistry* **36:** 7890–7897.

Borowski, P., Heiland, M., Feucht, H., Laufs, R. 1999, Characterisation of non-structural protein 3 of hepatitis C virus as modulator of protein phosphorylation mediated by PKA and PKC: evidences for action on the level of substrate and enzyme. *Arch Virol* **144:** 687–701.

Boulant, S., Montserret, R., Hope, R.G., Ratinier, M., Targett-Adams, P., Lavergne, J.P., Penin, F., McLauchlan, J. 2006, Structural determinants that target the hepatitis C virus core protein to lipid droplets. *J Biol Chem* **281:** 22236–22247.

Boulant, S., Targett-Adams, P., McLauchlan, J. 2007, Disrupting the association of hepatitis C virus core protein with lipid droplets correlates with a loss in production of infectious virus. *J Gen Virol* **88:** 2204–2213.

Breiman, A., Grandvaux, N., Lin, R., Ottone, C., Akira, S., Yoneyama, M., Fujita, T., Hiscott, J., Meurs, E.F. 2005, Inhibition of RIG-I-dependent signaling to the interferon pathway during hepatitis C virus expression and restoration of signaling by IKKepsilon. *J Virol* **79:** 3969–3978.

Bukh, J., Purcell, R.H., Miller, R.H. 1994, Sequence analysis of the core gene of 14 hepatitis C virus genotypes. *Proc Natl Acad Sci USA* **91:** 8239–8243.

Carroll, S.S., Olsen, D.B. 2006, Nucleoside analog inhibitors of hepatitis C virus replication. *Infect Disord Drug Targets* **6:** 17–29.

Chen, S.L., Morgan, T.R. 2006, The natural history of hepatitis C virus HCV infection. *Int J Med Sci* **3:** 47–52.

Chen, Z., Benureau, Y., Rijnbrand, R., Yi, J., Wang, T., Warter, L., Lanford, R.E., Weinman, S.A., Lemon, S.M., Martin, A., Li, K. 2007, GB virus B disrupts RIG-I signaling by NS3/4A-mediated cleavage of the adaptor protein MAVS. *J Virol* **81:** 964–976.

Choo, Q.L., Kuo, G., Weiner, A.J., Overby, L.R., Bradley, D.W., Houghton, M. 1989, Isolation of a cDNA clone derived from a blood-borne non-A, non-B viral hepatitis genome. *Science* **244:** 359–362.

Choo, Q.L., Richman, K.H., Han, J.H., Berger, K., Lee, C., Dong, C., Gallegos, C., Coit, D., Medina-Selby, R., Barr, P.J., et al. 1991, Genetic organization and diversity of the hepatitis C virus. *Proc Natl Acad Sci USA* **88:** 2451–2455.

Darke, P.L., Jacobs, A.R., Waxman, L., Kuo, L.C. 1999, Inhibition of hepatitis C virus NS2/3 processing by NS4A peptides. Implications for control of viral processing. *J Biol Chem* **274:** 34511–34514.

Day, C.L., Lauer, G.M., Robbins, G.K., McGovern, B., Wurcel, A.G., Gandhi, R.T., Chung, R.T., Walker, B.D. 2002, Broad specificity of virus-specific CD4+ T-helper-cell responses in resolved hepatitis C virus infection. *J Virol* **76:** 12584–12595.

De Francesco, R., Pessi, A., Steinkuhler, C. 1998, The hepatitis C virus NS3 proteinase: structure and function of a zinc-containing serine proteinase. *Antivir Ther* **3:** 99–109.

Dimitrova, M., Imbert, I., Kieny, M.P., Schuster, C. 2003, Protein–protein interactions between hepatitis C virus nonstructural proteins. *J Virol* **77:** 5401–5414.

Dumoulin, F.L., Von Dem Bussche, A., Li, J., Khamzina, L., Wands, J.R., Sauerbruch, T., Spengler, U. 2003, Hepatitis C virus NS2 protein inhibits gene expression from different cellular and viral promoters in hepatic and nonhepatic cell lines. *Virology* **305:** 260–266.

Erdtmann, L., Franck, N., Lerat, H., Le Seyec, J., Gilot, D., Cannie, I., Gripon, P., Hibner, U., Guguen-Guillouzo, C. 2003, The hepatitis C virus NS2 protein is an inhibitor of CIDE-B-induced apoptosis. *J Biol Chem* **278:** 18256–18264.

Failla, C., Tomei, L., De Francesco, R. 1994, Both NS3 and NS4A are required for proteolytic processing of hepatitis C virus nonstructural proteins. *J Virol* **68:** 3753–3760.

Flajolet, M., Rotondo, G., Daviet, L., Bergametti, F., Inchauspe, G., Tiollais, P., Transy, C., Legrain, P. 2000, A genomic approach of the hepatitis C virus generates a protein interaction map. *Gene* **242:** 369–379.

Foy, E., Li, K., Sumpter, R., Jr., Loo, Y.M., Johnson, C.L., Wang, C., Fish, P.M., Yoneyama, M., Fujita, T., Lemon, S.M., Gale, M. Jr. 2005, Control of antiviral defenses through hepatitis C virus disruption of retinoic acid-inducible gene-I signaling. *Proc Natl Acad Sci USA* **102:** 2986–2991.

Franck, N., Le Seyec, J., Guguen-Guillouzo, C., Erdtmann, L. 2005, Hepatitis C virus NS2 protein is phosphorylated by the protein kinase CK2 and targeted for degradation to the proteasome. *J. Virol.* **79:** 2700–2708.

Gallinari, P., Brennan, D., Nardi, C., Brunetti, M., Tomei, L., Steinkuhler, C., De Francesco, R. 1998, Multiple enzymatic activities associated with recombinant NS3 protein of hepatitis C virus. *J Virol* **72:** 6758–6769.

Gallinari, P., Paolini, C., Brennan, D., Nardi, C., Steinkuhler, C., De Francesco, R. 1999, Modulation of hepatitis C virus NS3 protease and helicase activities through the interaction with NS4A. *Biochemistry* **38:** 5620–5632.

Grakoui, A., McCourt, D.W., Wychowski, C., Feinstone, S.M., Rice, C.M. 1993a, Characterization of the hepatitis C virus-encoded serine proteinase: determination of proteinase-dependent polyprotein cleavage sites. *J Virol* **67:** 2832–2843.

Grakoui, A., McCourt, D.W., Wychowski, C., Feinstone, S.M., Rice, C.M. 1993b, A second hepatitis C virus-encoded proteinase. *Proc Natl Acad Sci USA* **90:** 10583–10587.

Grakoui, A., Shoukry, N.H., Woollard, D.J., Han, J.H., Hanson, H.L., Ghrayeb, J., Murthy, K.K., Rice, C.M., Walker, C.M. 2003, HCV persistence and immune evasion in the absence of memory T cell help. *Science* **302:** 659–662.

Gwack, Y., Kim, D.W., Han, J.H., Choe, J. 1997, DNA helicase activity of the hepatitis C virus nonstructural protein 3. *Eur J Biochem* **250:** 47–54.

Hadziyannis, S.J., Sette, H. Jr., Morgan, T.R., Balan, V., Diago, M., Marcellin, P., Ramadori, G., Bodenheimer, H. Jr., Bernstein, D., Rizzetto, M., Zeuzem, S., Pockros, P.J., Lin, A., Ackrill, A.M. 2004, Peginterferon-alpha2a and ribavirin combination therapy in chronic hepatitis C: a randomized study of treatment duration and ribavirin dose. *Ann Intern Med* **140:** 346–355.

Hassan, M., Ghozlan, H., Abdel-Kader, O. 2005, Activation of c-Jun NH2-terminal kinase JNK signaling pathway is essential for the stimulation of hepatitis C virus HCV non-structural protein 3 NS3-mediated cell growth. *Virology* **333:** 324–336.

Hijikata, M., Mizushima, H., Akagi, T., Mori, S., Kakiuchi, N., Kato, N., Tanaka, T., Kimura, K., Shimotohno, K. 1993, Two distinct proteinase activities required for the processing of a putative nonstructural precursor protein of hepatitis C virus. *J Virol* **67:** 4665–4675.

Hinrichsen, H., Benhamou, Y., Wedemeyer, H., Reiser, M., Sentjens, R.E., Calleja, J.L., Forns, X., Erhardt, A., Cronlein, J., Chaves, R.L., Yong, C.L., Nehmiz, G., Steinmann, G.G. 2004, Short-term antiviral efficacy of BILN 2061, a hepatitis C virus serine protease inhibitor, in hepatitis C genotype 1 patients. *Gastroenterology* **127:** 1347–1355.

Hoofnagle, J.H. 1997, Hepatitis C: the clinical spectrum of disease. *Hepatology* **26:** 15S–20S.

Hoofnagle, J.H. 2002, Course and outcome of hepatitis C. *Hepatology* **36:** S21–29.

Hope, R.G., McLauchlan, J. 2000, Sequence motifs required for lipid droplet association and protein stability are unique to the hepatitis C virus core protein. *J Gen Virol* **81:** 1913–1925.

Hope, V.D., Judd, A., Hickman, M., Lamagni, T., Hunter, G., Stimson, G.V., Jones, S., Donovan, L., Parry, J.V., Gill, O.N. 2001, Prevalence of hepatitis C among injection drug users in England and Wales: is harm reduction working? *Am J Public Health* **91:** 38–42.

Hughes, C.A., Shafran, S.D. 2006, Chronic hepatitis C virus management: 2000–2005 update. *Ann Pharmacother* **40:** 74–82.

Jirasko, V., Montserret, R., Appel, N., Janvier, A., Eustachi, L., Brohm, C., Steinmann, E., Pietschmann, T., Penin, F., Bartenschlager, R. 2008, Structural and functional characterization of non-structural protein 2 for its role in hepatitis C virus assembly. *J Biol Chem* **283:** 28546–28562.

Jones, C.T., Murray, C.L., Eastman, D.K., Tassello, J., Rice, C.M. 2007, Hepatitis C virus p7 and NS2 proteins are essential for infectious virus production. *J Virol* **81:** 8374–8383.

Kakiuchi, N., Nishikawa, S., Hattori, M., Shimotohno, K. 1999, A high throughput assay of the hepatitis C virus nonstructural protein 3 serine proteinase. *J Virol Meth* **80:** 77–84.

Karayiannis, P., McGarvey, M.J. 1995, The GB hepatitis viruses. *J Viral Hepat* **2:** 221–226.

Kato, T., Date, T., Miyamoto, M., Furusaka, A., Tokushige, K., Mizokami, M., Wakita, T. 2003, Efficient replication of the genotype 2a hepatitis C virus subgenomic replicon. *Gastroenterology* **125:** 1808–1817.

Kaukinen, P., Sillanpaa, M., Kotenko, S., Lin, R., Hiscott, J., Melen, K., Julkunen, I. 2006, Hepatitis C virus NS2 and NS3/4A proteins are potent inhibitors of host cell cytokine/chemokine gene expression. *Virol J* **3:** 66.

Kiiver, K., Merits, A., Ustav, M., Zusinaite, E. 2006, Complex formation between hepatitis C virus NS2 and NS3 proteins. *Virus Res* **117:** 264–272.

Kim, D.W., Gwack, Y., Han, J.H., Choe, J. 1995, C-terminal domain of the hepatitis C virus NS3 protein contains an RNA helicase activity. *Biochem Biophys Res Commun* **215:** 160–166.

Kim, J.E., Song, W.K., Chung, K.M., Back, S.H., Jang, S.K. 1999, Subcellular localization of hepatitis C viral proteins in mammalian cells. *Arch Virol* **144:** 329–343.

Kim, J.L., Morgenstern, K.A., Lin, C., Fox, T., Dwyer, M.D., Landro, J.A., Chambers, S.P., Markland, W., Lepre, C.A., O'Malley, E.T., Harbeson, S.L., Rice, C.M., Murcko, M.A., Caron, P.R., Thomson, J.A. 1996, Crystal structure of the hepatitis C virus NS3 protease domain complexed with a synthetic NS4A cofactor peptide. *Cell* **87:** 343–355.

Kim, K.M., Kwon, S.N., Kang, J.I., Lee, S.H., Jang, S.K., Ahn, B.Y., Kim, Y.K. 2007, Hepatitis C virus NS2 protein activates cellular cyclic AMP-dependent pathways. *Biochem Biophys Res Commun* **356:** 948–954.

Krieger, N., Lohmann, V., Bartenschlager, R. 2001, Enhancement of hepatitis C virus RNA replication by cell culture-adaptive mutations. *J Virol* **75:** 4614–4624.

Kuo, G., Choo, Q.L., Alter, H.J., Gitnick, G.L., Redeker, A.G., Purcell, R.H., Miyamura, T., Dienstag, J.L., Alter, M.J., Stevens, C.E., et al. 1989, An assay for circulating antibodies to a major etiologic virus of human non-A, non-B hepatitis. *Science* **244:** 362–364.

Lackner, T., Muller, A., Pankraz, A., Becher, P., Thiel, H.J., Gorbalenya, A.E., Tautz, N. 2004, Temporal modulation of an autoprotease is crucial for replication and pathogenicity of an RNA virus. *J Virol* **78:** 10765–10775.

Lamarre, D., Anderson, P.C., Bailey, M., Beaulieu, P., Bolger, G., Bonneau, P., Bos, M., Cameron, D.R., Cartier, M., Cordingley, M.G., Faucher, A.M., Goudreau, N., Kawai, S.H., Kukolj, G., Lagace, L., LaPlante, S.R., Narjes, H., Poupart, M.A., Rancourt, J., Sentjens, R.E.,

St George, R., Simoneau, B., Steinmann, G., Thibeault, D., Tsantrizos, Y.S., Weldon, S.M., Yong, C.L., Llinas-Brunet, M. 2003, An NS3 protease inhibitor with antiviral effects in humans infected with hepatitis C virus. *Nature* **426:** 186–189.

Li, K., Chen, Z., Kato, N., Gale, M. Jr., Lemon, S.M. 2005a, Distinct polyI-C and virus-activated signaling pathways leading to interferon-beta production in hepatocytes. *J Biol Chem* **280:** 16739–16747.

Li, X.D., Sun, L., Seth, R.B., Pineda, G., Chen, Z.J. 2005b, Hepatitis C virus protease NS3/4A cleaves mitochondrial antiviral signaling protein off the mitochondria to evade innate immunity. *Proc Natl Acad Sci USA* **102:** 17717–17722.

Lin, C., Pragai, B.M., Grakoui, A., Xu, J., Rice, C.M. 1994, Hepatitis C virus NS3 serine proteinase: trans-cleavage requirements and processing kinetics. *J Virol* **68:** 8147–8157.

Lin, C., Thomson, J.A., Rice, C.M. 1995, A central region in the hepatitis C virus NS4A protein allows formation of an active NS3-NS4A serine proteinase complex in vivo and in vitro. *J Virol* **69:** 4373–4380.

Lindenbach, B.D., Rice, C.M. 2001, Flaviviridae: the viruses and their replication. In *Field's Virology*. D.M. Knipe, P.M. Howley (Eds.): Lippincott Williams & Wilkins, Philadelphia, PA.

Lindenbach, B.D., Rice, C.M. 2005, Unravelling hepatitis C virus replication from genome to function. *Nature* **436:** 933–938.

Lindenbach, B.D., Evans, M.J., Syder, A.J., Wolk, B., Tellinghuisen, T.L., Liu, C.C., Maruyama, T., Hynes, R.O., Burton, D.R., McKeating, J.A., Rice, C.M. 2005, Complete replication of hepatitis C virus in cell culture. *Science* **309:** 623–626.

Lo, S.Y., Selby, M., Tong, M., Ou, J.H. 1994, Comparative studies of the core gene products of two different hepatitis C virus isolates: two alternative forms determined by a single amino acid substitution. *Virology* **199:** 124–131.

Lohmann, V., Korner, F., Koch, J., Herian, U., Theilmann, L., Bartenschlager, R. 1999, Replication of subgenomic hepatitis C virus RNAs in a hepatoma cell line. *Science* **285:** 110–113.

Lopez-Rovira, T., Chalaux, E., Rosa, J.L., Bartrons, R., Ventura, F. 2000, Interaction and functional cooperation of NF-kappa B with smads. Transcriptional regulation of the junB promoter. *J Biol Chem* **275:** 28937–28946.

Lorenz, I.C., Marcotrigiano, J., Dentzer, T.G., Rice, C.M. 2006, Structure of the catalytic domain of the hepatitis C virus NS2–3 protease. *Nature* **442:** 831–835.

Love, R.A., Parge, H.E., Wickersham, J.A., Hostomsky, Z., Habuka, N., Moomaw, E.W., Adachi, T., Hostomska, Z. 1996, The crystal structure of hepatitis C virus NS3 proteinase reveals a trypsin-like fold and a structural zinc binding site. *Cell* **87:** 331–342.

Love, R.A., Parge, H.E., Wickersham, J.A., Hostomsky, Z., Habuka, N., Moomaw, E.W., Adachi, T., Margosiak, S., Dagostino, E., Hostomska, Z. 1998, The conformation of hepatitis C virus NS3 proteinase with and without NS4A: a structural basis for the activation of the enzyme by its cofactor. *Clin Diagn Virol* **10:** 151–156.

Major, M.E., Rehermann, B., Feinstone, S. 2001, Hepatitis C Virus. In *Field's Virology*. D.M. Knipe, P.M. Howley (Eds.): Lippincott Williams & Wilkins, Philadelphia, PA.

McLauchlan, J., Lemberg, M.K., Hope, G., Martoglio, B. 2002, Intramembrane proteolysis promotes trafficking of hepatitis C virus core protein to lipid droplets. *Embo J* **21:** 3980–3988.

Moradpour, D., Englert, C., Wakita, T., Wands, J.R. 1996, Characterization of cell lines allowing tightly regulated expression of hepatitis C virus core protein. *Virology* **222:** 51–63.

Moradpour, D., Penin, F., Rice, C.M. 2007, Replication of hepatitis C virus. *Nat Rev Microbiol* **5:** 453–463.

Mottola, G., Cardinali, G., Ceccacci, A., Trozzi, C., Bartholomew, L., Torrisi, M.R., Pedrazzini, E., Bonatti, S., Migliaccio, G. 2002, Hepatitis C virus nonstructural proteins are localized in a modified endoplasmic reticulum of cells expressing viral subgenomic replicons. *Virology* **293:** 31–43.

Murayama, A., Date, T., Morikawa, K., Akazawa, D., Miyamoto, M., Kaga, M., Ishii, K., Suzuki, T., Kato, T., Mizokami, M., Wakita, T. 2007, The NS3 helicase and NS5B-to-3′X regions are important for efficient hepatitis C virus strain JFH-1 replication in Huh7 cells. *J Virol* **81:** 8030–8040.

Ogata, N., Alter, H.J., Miller, R.H., Purcell, R.H. 1991, Nucleotide sequence and mutation rate of the H strain of hepatitis C virus. *Proc Natl Acad Sci USA* **88:** 3392–3396.

Pallaoro, M., Lahm, A., Biasiol, G., Brunetti, M., Nardella, C., Orsatti, L., Bonelli, F., Orru, S., Narjes, F., Steinkuhler, C. 2001, Characterization of the hepatitis C virus NS2/3 processing reaction by using a purified precursor protein. *J Virol* **75:** 9939–9946.

Pieroni, L., Santolini, E., Fipaldini, C., Pacini, L., Migliaccio, G., La Monica, N. 1997, In vitro study of the NS2–3 protease of hepatitis C virus. *J Virol* **71:** 6373–6380.

Pietschmann, T., Kaul, A., Koutsoudakis, G., Shavinskaya, A., Kallis, S., Steinmann, E., Abid, K., Negro, F., Dreux, M., Cosset, F.L., Bartenschlager, R. 2006, Construction and characterization of infectious intragenotypic and intergenotypic hepatitis C virus chimeras. *Proc Natl Acad Sci USA* **103:** 7408–7413.

Prikhod'ko, E.A., Prikhod'ko, G.G., Siegel, R.M., Thompson, P., Major, M.E., Cohen, J.I. 2004, The NS3 protein of hepatitis C virus induces caspase-8-mediated apoptosis independent of its protease or helicase activities. *Virology* **329:** 53–67.

Racanelli, V., Manigold, T. 2007, Presentation of HCV antigens to naive CD8+T cells: why the where, when, what and how are important for virus control and infection outcome. *Clin Immunol* **124:** 5–12.

Reed, K.E., Grakoui, A., Rice, C.M. 1995, Hepatitis C virus-encoded NS2-3 protease: cleavage-site mutagenesis and requirements for bimolecular cleavage. *J Virol* **69:** 4127–4136.

Reesink, H.W., Zeuzem, S., Weegink, C.J., Forestier, N., Van Vliet, A., Van de Wetering de Rooij, J., McNair, L., Purdy, S., Kauffman, R., Alam, J., Jansen, P.L.M. 2006, Rapid decline of viral RNA in hepatitis C patients treated with VX-950: a phase Ib, placebo-controlled, randomized study. *Gastroenterology* **131:** 997.

Reiser, M., Hinrichsen, H., Benhamou, Y., Reesink, H.W., Wedemeyer, H., Avendano, C., Riba, N., Yong, C.L., Nehmiz, G., Steinmann, G.G. 2005, Antiviral efficacy of NS3-serine protease inhibitor BILN-2061 in patients with chronic genotype 2 and 3 hepatitis C. *Hepatology* **41:** 832–835.

Rinck, G., Birghan, C., Harada, T., Meyers, G., Thiel, H.J., Tautz, N. 2001, A cellular J-domain protein modulates polyprotein processing and cytopathogenicity of a pestivirus. *J Virol* **75:** 9470–9482.

Rouille, Y., Helle, F., Delgrange, D., Roingeard, P., Voisset, C., Blanchard, E., Belouzard, S., McKeating, J., Patel, A.H., Maertens, G., Wakita, T., Wychowski, C., Dubuisson, J. 2006, Subcellular localization of hepatitis C virus structural proteins in a cell culture system that efficiently replicates the virus. *J Virol* **80:** 2832–2841.

Rubbia-Brandt, L., Quadri, R., Abid, K., Giostra, E., Male, P.J., Mentha, G., Spahr, L., Zarski, J.P., Borisch, B., Hadengue, A., Negro, F. 2000, Hepatocyte steatosis is a cytopathic effect of hepatitis C virus genotype 3. *J Hepatol* **33:** 106–115.

Saito, T., Owen, D.M., Jiang, F., Marcotrigiano, J., Gale, M. Jr. 2008, Innate immunity induced by composition-dependent RIG-I recognition of hepatitis C virus RNA. *Nature* **454:** 523–527.

Sakamuro, D., Furukawa, T., Takegami, T. 1995, Hepatitis C virus nonstructural protein NS3 transforms NIH 3T3 cells. *J Virol* **69:** 3893–3896.

Sarrazin, C., Rouzier, R., Wagner, F., Forestier, N., Larrey, D., Gupta, S.K., Hussain, M., Shah, A., Cutler, D., Zhang, J., Zeuzem, S. 2007, SCH 503034, a novel hepatitis C virus protease inhibitor, plus pegylated interferon alpha-2b for genotype 1 nonresponders. *Gastroenterology* **132:** 1270–1278.

Seiwert, S., Hong, S., Lim, S., Tan, H., Kossen, K., Blatt, L. 2006, Abstract 2. In *First International Workshop on Hepatitis C: Resistance and New Compounds*. Boston, MA.

Shi, S.T., Lee, K.J., Aizaki, H., Hwang, S.B., Lai, M.M. 2003, Hepatitis C virus RNA replication occurs on a detergent-resistant membrane that cofractionates with caveolin-2. *J Virol* **77:** 4160–4168.

Simmonds, P. 2004, Genetic diversity and evolution of hepatitis C virus-15 years on. *J Gen Virol* **85:** 3173–3188.

Simmonds, P., Bukh, J., Combet, C., Deleage, G., Enomoto, N., Feinstone, S., Halfon, P., Inchauspe, G., Kuiken, C., Maertens, G., Mizokami, M., Murphy, D.G., Okamoto, H.,

Pawlotsky, J.M., Penin, F., Sablon, E., Shin, I.T., Stuyver, L.J., Thiel, H.J., Viazov, S., Weiner, A.J., Widell, A. 2005, Consensus proposals for a unified system of nomenclature of hepatitis C virus genotypes. *Hepatology* **42:** 962–973.

Simons, J.N., Leary, T.P., Dawson, G.J., Pilot-Matias, T.J., Muerhoff, A.S., Schlauder, G.G., Desai, S.M., Mushahwar, I.K. 1995a, Isolation of novel virus-like sequences associated with human hepatitis. *Nat Med* **1:** 564–569.

Simons, J.N., Pilot-Matias, T.J., Leary, T.P., Dawson, G.J., Desai, S.M., Schlauder, G.G., Muerhoff, A.S., Erker, J.C., Buijk, S.L., Chalmers, M.L., et al. 1995b, Identification of two flavivirus-like genomes in the GB hepatitis agent. *Proc Natl Acad Sci USA* **92:** 3401–3405.

Sommergruber, W., Casari, G., Fessl, F., Seipelt, J., Skern, T. 1994, The 2A proteinase of human rhinovirus is a zinc containing enzyme. *Virology* **204:** 815–818.

Steinkuhler, C., Urbani, A., Tomei, L., Biasiol, G., Sardana, M., Bianchi, E., Pessi, A., De Francesco, R. 1996, Activity of purified hepatitis C virus protease NS3 on peptide substrates. *J Virol* **70:** 6694–6700.

Stempniak, M., Hostomska, Z., Nodes, B.R., Hostomsky, Z. 1997, The NS3 proteinase domain of hepatitis C virus is a zinc-containing enzyme. *J Virol* **71:** 2881–2886.

Sumpter, R. Jr., Loo, Y.M., Foy, E., Li, K., Yoneyama, M., Fujita, T., Lemon, S.M., Gale, M. Jr. 2005, Regulating intracellular antiviral defense and permissiveness to hepatitis C virus RNA replication through a cellular RNA helicase, RIG-I. *J Virol* **79:** 2689–2699.

Suzich, J.A., Tamura, J.K., Palmer-Hill, F., Warrener, P., Grakoui, A., Rice, C.M., Feinstone, S.M., Collett, M.S. 1993, Hepatitis C virus NS3 protein polynucleotide-stimulated nucleoside triphosphatase and comparison with the related pestivirus and flavivirus enzymes. *J Virol* **67:** 6152–6158.

Tedbury, P.R., Harris, M. 2007, Characterisation of the role of zinc in the hepatitis C virus NS2/3 auto-cleavage and NS3 protease activities. *J Mol Biol* **366:** 1652–1660.

The Global Burden of Hepatitis C Working Group. 2004, Global burden of disease GBD for hepatitis C. *J Clin Pharmacol* **44:** 20–29.

Thibeault, D., Maurice, R., Pilote, L., Lamarre, D., Pause, A. 2001, In vitro characterization of a purified NS2/3 protease variant of hepatitis C virus. *J Biol Chem* **276:** 46678–46684.

Thimme, R., Bukh, J., Spangenberg, H.C., Wieland, S., Pemberton, J., Steiger, C., Govindarajan, S., Purcell, R.H., Chisari, F.V. 2002, Viral and immunological determinants of hepatitis C virus clearance, persistence, and disease. *Proc Natl Acad Sci USA* **99:** 15661–15668.

Tomei, L., Failla, C., Santolini, E., De Francesco, R., La Monica, N. 1993, NS3 is a serine protease required for processing of hepatitis C virus polyprotein. *J Virol* **67:** 4017–4026.

Vauloup-Fellous, C., Pene, V., Garaud-Aunis, J., Harper, F., Bardin, S., Suire, Y., Pichard, E., Schmitt, A., Sogni, P., Pierron, G., Briand, P., Rosenberg, A.R. 2006, Signal peptide peptidase-catalyzed cleavage of hepatitis C virus core protein is dispensable for virus budding but destabilizes the viral capsid. *J Biol Chem* **281:** 27679–27692.

Wakita, T., Pietschmann, T., Kato, T., Date, T., Miyamoto, M., Zhao, Z., Murthy, K., Habermann, A., Krausslich, H.G., Mizokami, M., Bartenschlager, R., Liang, T.J. 2005, Production of infectious hepatitis C virus in tissue culture from a cloned viral genome. *Nat Med* **11:** 791–796.

Waxman, L., Whitney, M., Pollok, B.A., Kuo, L.C., Darke, P.L. 2001, Host cell factor requirement for hepatitis C virus enzyme maturation. *Proc Natl Acad Sci USA* **98:** 13931–13935.

Wedemeyer, H., He, X.S., Nascimbeni, M., Davis, A.R., Greenberg, H.B., Hoofnagle, J.H., Liang, T.J., Alter, H., Rehermann, B. 2002, Impaired effector function of hepatitis C virus-specific CD8+ T cells in chronic hepatitis C virus infection. *J Immunol* **169:** 3447–3458.

Weigand, K., Stremmel, W., Encke, J. 2007, Treatment of hepatitis C virus infection. *World J Gastroenterol* **13:** 1897–1905.

Welbourn, S., Green, R., Gamache, I., Dandache, S., Lohmann, V., Bartenschlager, R., Meerovitch, K., Pause, A. 2005, Hepatitis C virus NS2/3 processing is required for NS3 stability and viral RNA replication. *J Biol Chem* **280:** 29604–29611.

Whitney, M., Stack, J.H., Darke, P.L., Zheng, W., Terzo, J., Inglese, J., Strulovici, B., Kuo, L.C., Pollock, B.A. 2002, A collaborative screening program for the discovery of inhibitors of HCV NS2/3 cis-cleaving protease activity. *J Biomol Screen* **7:** 149–154.

Yang, W., Zhao, Y., Fabrycki, J., Hou, X., Nie, X., Sanchez, A., Phadke, A., Deshpande, M., Agarwal, A., Huang, M. 2008, Selection of replicon variants resistant to ACH-806, a novel hepatitis C virus inhibitor with no cross-resistance to NS3 protease and NS5B polymerase inhibitors. *Antimicrob Agents Chemother* **52:** 2043–2052.

Yang, X.J., Liu, J., Ye, L., Liao, Q.J., Wu, J.G., Gao, J.R., She, Y.L., Wu, Z.H., Ye, L.B. 2006, HCV NS2 protein inhibits cell proliferation and induces cell cycle arrest in the S-phase in mammalian cells through down-regulation of cyclin A expression. *Virus Res* **121:** 134–143.

Yasui, K., Wakita, T., Tsukiyama-Kohara, K., Funahashi, S.I., Ichikawa, M., Kajita, T., Moradpour, D., Wands, J.R., Kohara, M. 1998, The native form and maturation process of hepatitis C virus core protein. *J Virol* **72:** 6048–6055.

Yokosuka, O., Kojima, H., Imazeki, F., Tagawa, M., Saisho, H., Tamatsukuri, S., Omata, M. 1999, Spontaneous negativation of serum hepatitis C virus RNA is a rare event in type C chronic liver diseases: analysis of HCV RNA in 320 patients who were followed for more than 3 years. *J Hepatol* **31:** 394–399.

Yu, S.F., Lloyd, R.E. 1992, Characterization of the roles of conserved cysteine and histidine residues in poliovirus 2A protease. *Virology* **186:** 725–735.

Zemel, R., Gerechet, S., Greif, H., Bachmatove, L., Birk, Y., Golan-Goldhirsh, A., Kunin, M., Berdichevsky, Y., Benhar, I., Tur-Kaspa, R. 2001, Cell transformation induced by hepatitis C virus NS3 serine protease. *J Viral Hepat* **8:** 96–102.

Zhong, J., Gastaminza, P., Cheng, G., Kapadia, S., Kato, T., Burton, D.R., Wieland, S.F., Uprichard, S.L., Wakita, T., Chisari, F.V. 2005, Robust hepatitis C virus infection in vitro. *Proc Natl Acad Sci USA* **102:** 9294–9299.

Chapter 4
Antiviral Activity of Proteasome Inhibitors/Cytomegalovirus

Marion Kaspari and Elke Bogner

Abstract Human cytomegalovirus (HCMV) is a member of the herpesvirus family and represents a major human pathogen causing life-threatening diseases. HCMV persists lifelong in its host because the immune system does not achieve clearance of the virus. Since the currently available anti-HCMV drugs cause multiple problems new antiviral therapeutics are urgently in demand. The ubiquitin-proteasome system (UPS) not only mediate the degradation of mis-folded proteins but is also involved in processes essential for viral replication. Thus the UPS could represent a new target for anti-HCMV therapy supporting a different mode of action. In this chapter we will review the impact of proteasome inhibition on HCMV replication and discuss whether the use of proteasome inhibitors is a possible strategy to prevent viral replication.

Keywords human cytomegalovirus • viral replication • proteasome inhibitors • antiviral therapy

4.1 Introduction

Human cytomegalovirus (HCMV), one of eight human herpesviruses, can cause serious, life threatening diseases in newborns and immunocompromised patients, i.e. organ transplant recipients and patients with AIDS. HCMV is widespread throughout the population world wide with a seroprevalence of up to 100% in adults. To date nearly all available anti-HCMV drugs are inhibitors of the viral DNA polymerase. Due to multiple problems caused by the current available drugs development of new antiviral compounds which are non-nucleosidic and have a different mode of action is needed.

A new target for antiviral therapy is viral maturation because the involved virus-specific processes are absent in host cells. HCMV maturational events

M. Kaspari and E. Bogner (✉)
Charité University Hospital, Berlin, Institute of Virology, Helmut-Ruska-Haus, Chariteplatz 1, 10117 Berlin, Germany
e-mail: elke.bogner@charite.de

U. Lendeckel and N.M. Hooper (eds.), *Viral Proteases and Antiviral Protease Inhibitor Therapy*, Proteases in Biology and Disease 8,

of DNA replication and capsid assembly require efficient translocation of viral gene products into the nucleus of infected cells. Several gene products of HCMV have been identified that appear to be involved in the process of DNA replication, cleavage and packaging (Penfold and Mocarski, 1997; Gibson, 1996).

In addition, HCMV like many other viruses has evolved strategies to redistribute host proteins and to take over host functions to promote viral replication. One mechanism is the inhibition of viral antigen presentation via MHC class I molecules. Multiple ways of inhibition have been established by HCMV immune evasion proteins that block different steps of folding and assembly pathways of MHC class I molecules. While the viral protein US6 prevents peptide translocation into the ER, US3 leads to retention of MHC class I complexes in the endoplasmic reticulum (ER) and US2 as well as US11 mediate retrograde transport of MHC class I heavy chains from the ER to the cytoplasm (Hengel et al., 1997; Hewitt et al., 2001; Momburg and Hengel, 2001; Wiertz et al., 1996). The latter are modified by polyubiquitinylation leading to degradation by the proteasome. The ubiquitin-proteasome system (UPS) mediates degradation of misfolded proteins but it is also required for specific processing events (Ciechanover, 1998; Coux et al., 1996). Besides regulated protein turnover the proteasome is involved in apoptosis, cell cycle, protein sorting and regulation of transcription and signaling. Since these processes interfere with viral replication, proteasome inhibitors are now in focus as new targets for antiviral therapy.

4.2 Proteasome and Inhibitors

4.2.1 Proteasome

Proteasomes are enzymes of the non-lysosomal pathway of protein degradation in cells of eukaryotes, archaea and some bacteria. They were originally discovered as a large complex that degrades proteins conjugated to ubiquitin (Ciechanover, 1994). Their purpose is to degrade misfolded and short-lived proteins by a chemical break of peptide bonds. In addition proteasomes are involved in all ubiquitin-related processes. These include stress response, cell-cycle regulation, cell differentiation and antigen presentation. Proteasomes exist in two forms: a 20S and a 26S complex. The eukaryotic 20S proteasome consists of seven different α subunits and seven different β subunits, organized as α-β-β-α rings. The 26S complex is yielded by ATP-dependent conversion of the 20S form. In eukaryotes the 19S is like a cap located on one or both ends of the cylinder-shaped 20S proteasome (Fig. 4.1). While the 20S proteasome has protease activity the 19S cap represents a regulatory subunit for binding to ubiquitin conjugates (Coux et al., 1996). The enzymatically active sites are inside the cylinder and only unfolded, poly-ubiquitinated proteins can pass through the 19S cap into the cylinder.

4.2.2 *Inhibitors*

Based on the unique proteolytic mechanism of the proteasome it was possible to synthesize specific proteasome inhibitors (PI). The first inhibitors were peptide aldehydes that interact with the catalytic threonine. Numerous proteasome inhibitors are known, some of synthetic (like MG132) or natural origin (lactacystin). PIs inhibit NF-κB activation by preventing IκB degradation. To date reports are available suggesting proteasome inhibitors as a potential treatment for various diseases.

4.2.2.1 MG132

The PI MG132, a peptide aldehyde (Fig. 4.2), inhibits the chymotrypsin-like activity in a potent but reversible manner (Lee and Goldberg, 1998). In addition, it blocks degradation of short-lived proteins. Overall this peptide aldehyde is cell-permeable and represents a potent inhibitor of the proteasome.

4.2.2.2 Lactacystin

Lactacystin is a bacterial metabolite synthesized by bacteria of the genus *Streptomyces lactacystinaeus*. Lactacystin is a lactam or cyclic amid that inhibits the proteolytic activity of the proteasome (Fig. 4.3, Fenteany et al., 1995). The block of proteasome activity is mediated by covalent binding to the amino terminal threonine in the catalytic beta-subunit (Tomoda and Omura, 2000).

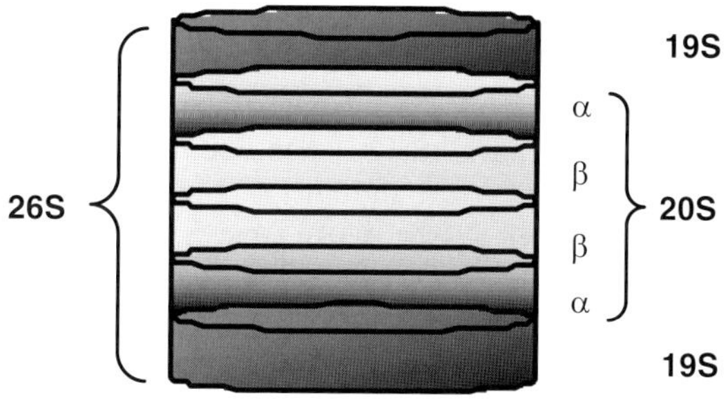

Fig. 4.1 Structure of the 26S proteasome. The 20S proteasome forms a cylinder with α and β subunits. The 20S complex is usually flanked by two 19S regulatory units

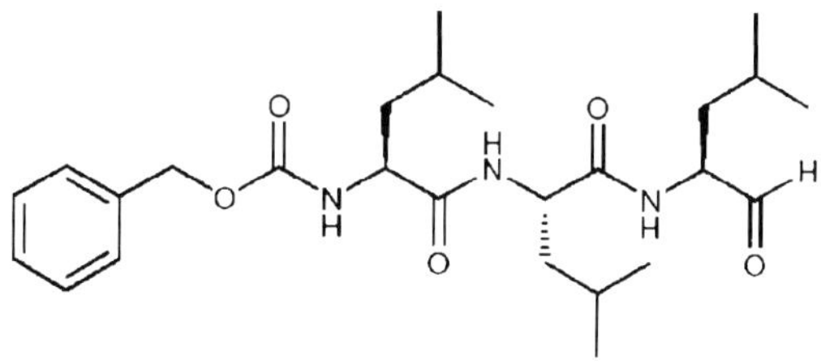

Fig. 4.2 Structure of MG132

Fig. 4.3 Structure of lactacystin

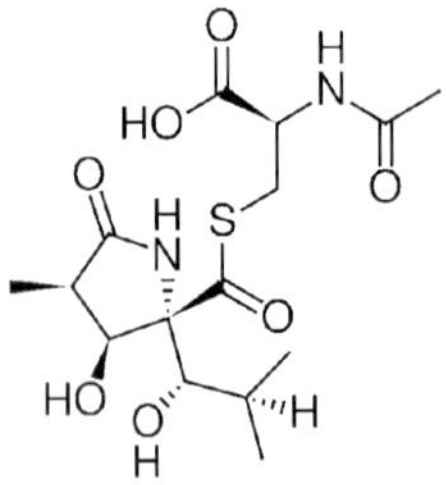

Another study demonstrated that the formation of intermediate structures plays an important role in the mode of action (Dick et al., 1997). The efficiency of lactacystin is dependent on its lactonization in the cell, generating β-lactone because the cells are only permeable to β-lactone. In the cell this substance has three different modes to react: (i) hydrolization into the inactive dihydroxy acid, (ii) formation of a sulfhydryl glutathione-conjugate (lactathionine) with analogous functions to lactacystin, or (iii) interaction with the proteasome (Dick et al., 1997). The group of Dick et al. demonstrated that lactathionine must undergo lactonization to yield β-lactone and serves as a reservoir for β-lactone in the cell. Until now, its complex synthesis prevents the development for therapeutical use.

4.3 Influence of Proteasome Inhibitors on HCMV Replication

4.3.1 HCMV Replication

HCMV has a sequential regulation of viral gene expression, leading to induction and repression cycles occurring in the immediate early (IE), early (E) and late phase (L) of viral replication. IE1 and IE2 induce expression of early proteins, mediate G1/S cell cycle arrest and host replication shut-off (Dittmer and Mocarski, 1997; Bresnahan et al., 1996). Early genes predominantly encode viral DNA replication factors, repair enzymes or immune evasion proteins and induce the expression of late genes (Mocarski and Courcelle, 2001). Late proteins down-regulate expression of early genes and are mainly involved in virion assembly (Fig. 4.3). HCMV DNA replication is thus the key step in maturation of new virions.

DNA replication occurs by rolling-circle mechanism and leads to the formation of head-to-tail linked concatemeric DNA which must be resolved into unit-length genomes during packaging. Proteins involved in viral DNA cleavage and packaging are so-called terminases. Terminases were first described as specific enzymatic proteins for dsDNA bacteriophages with multiple activities required for the whole packaging process (Black, 1988; Bhattacharyya and Rao, 1993; Feiss and Becker, 1983). Terminases are heteromultimers composed of a small subunit that binds DNA and a large subunit that harbors an endonuclease activity as well as

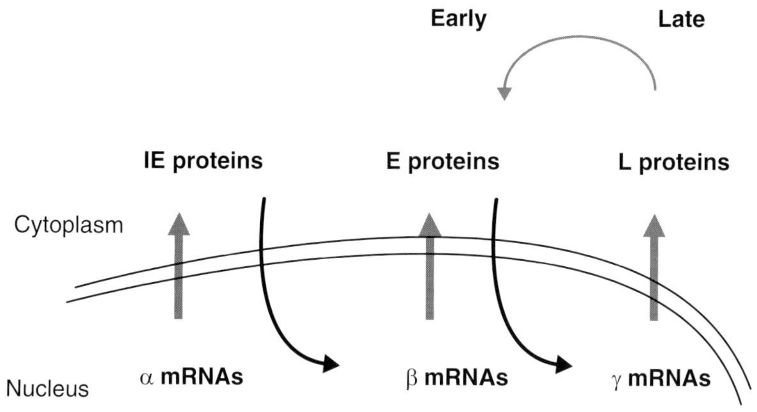

Fig. 4.4 Sequential viral gene expression of HCMV. IE proteins induce the expression of E genes. E genes are essential for replication and induce the transcription of L genes. L proteins are structural components of the virus and regulate the expression of E genes

an ATPase activity mediating the translocation of DNA into the preformed capsid (Becker and Murialdo, 1990; Bhattacharyya and Rao, 1994; Catalano et al., 1995; Fujisawa and Hearing, 1994; Feiss, 1986; Guo et al., 1987; Hang et al., 2000; Morita et al., 1993; Rao and Black, 1988).

We identified the terminase of human cytomegalovirus and demonstrated that it consists of the highly conserved proteins pUL56 and pUL89 (Bogner et al., 1993, 1998; Bogner, 2002; Hwang and Bogner, 2002; Scheffczik et al., 2002; Scholz et al., 2003; Thoma et al., 2006). While it is suggested that the large terminase subunit pUL56 (i) mediates the specific binding to packaging elements, so-called pac motifs, on the concatemers, (ii) provides energy for translocation of the DNA into preformed procapsids, and (iii) associates with the portal protein to enable the entry of the DNA, the small subunit pUL89 seems to be mainly required for completion of the packaging process by double nicking of concatemeric DNA into unit-length genomes (Scheffczik et al., 2002). Finally the capsid egress from the nucleus into the cytoplasm is mediated by budding through all nuclear membranes (Mettenleiter, 2002; Roizman and Knipe, 2001). The released naked nucleocapsids then traffic to the trans-Golgi network to acquire their final envelope (Fig. 4.5; Radsak et al., 1996).

4.3.2 Effects of Proteasome Inhibitors

Human cytomegalovirus like all known viruses essentially depends on the host cell metabolism and has evolved multiple strategies to take over host cell pathways and promote viral replication. The ubiquitin-proteasome system (UPS) seems to be one of the cellular pathways that viruses utilize for their own benefit. A number of studies have reported that proteasome inhibitors block replication of various viruses by targeting different steps of the replication cycle. For example, it has been demonstrated that proteasome inhibitors prevent translocation of incoming influenza virus to the nucleus (Khor et al., 2003). In the case of herpes simplex virus type 1 (HSV-1), proteasome inhibitors decreased immediate early and late viral protein expression (LaFrazia et al., 2006). Watanabe et al. (2005) reported that paramyxovirus maturation is blocked by proteasome inhibitors, while the UPS seems to be involved in budding events of rhabdoviruses (Harty et al., 2001) and HIV (Schubert et al., 2000).

Since it was known that PIs interfere with the viral replication cycle we addressed the question which steps in viral replication are affected. For our analyses we used the peptide aldehyde MG132. The antiviral activity of MG132 measured as the 50% inhibitory concentration (IC_{50}) against the laboratory strain HCMV AD169 was 0.16 μM in contrast to lactacystin with an IC_{50} of 13.53 μM.

The first step in the viral life-cycle is the entry into the host cell. By performing preadsorption and adsorption analysis in the presence and absence of MG132 we found that PIs have no effect on viral entry (Fig. 4.6). Experiments concerning the stage of viral replication that is sensitive to PI demonstrated that all phases

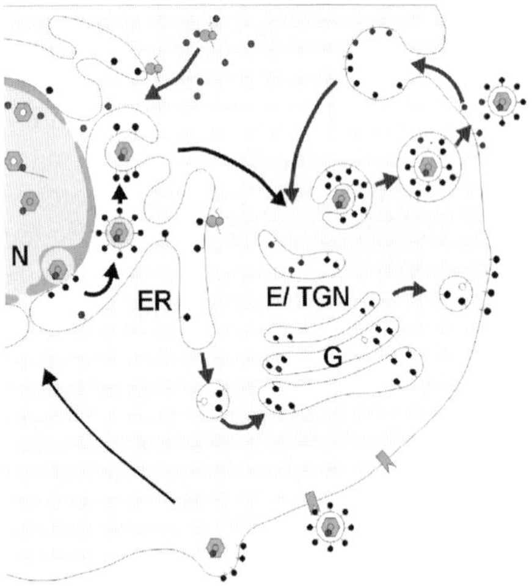

Fig. 4.5 HCMV assembly in infected cells. After penetration the DNA is released into the nucleus. Several E and L proteins are synthesized in the smooth ER and trans-located into the nucleus. At the late stage of infection capsids are formed, filled with DNA and bud into the cytoplasm. Final envelopment occurs at the TGN prior to release

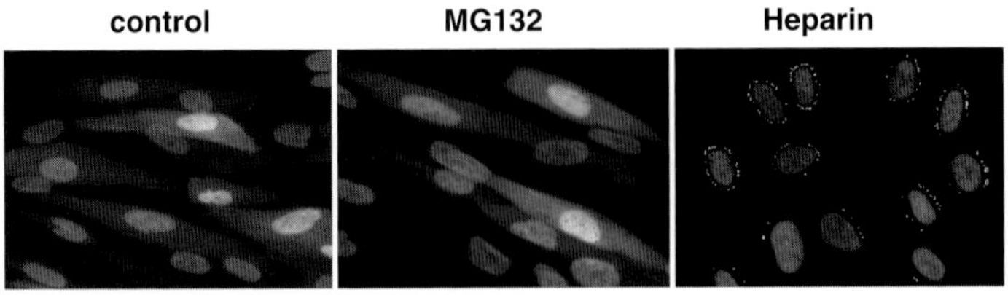

Fig. 4.6 Effect of MG132 on viral entry. Human fibroblasts were incubated with 0.5 μM MG132 or 1 mg/ml Heparin 30 min prior to infection with GFP-HCMV and during adsorption of the virus for 60 min. 24 h post infection immunofluorescence was performed with counterstaining of cell nuclei using Dapi. Untreated GFP-HCMV infected cells served as control

of viral replication are sensitive to MG132 treatment. The most noticeable effect was observed when the inhibitor was added in the IE or E phase, while addition of MG132 at late times post infection reduced viral replication by approximately 50%. Thus, proteasome activity seems to be required at multiple steps in HCMV replication (Kaspari et al., 2008).

An important step during viral maturation is the formation of infectious particles. Electron microscopy of infected cells in the presence of MG132 revealed that the formation of viral particles is completely inhibited (Fig. 4.7). Interestingly not only formation of infectious virions but also of non-infectious particles and even capsids was completely abolished in PI-treated cells (Fig. 4.7). These experiments indicated that MG132 leads to a complete breakdown of viral maturation.

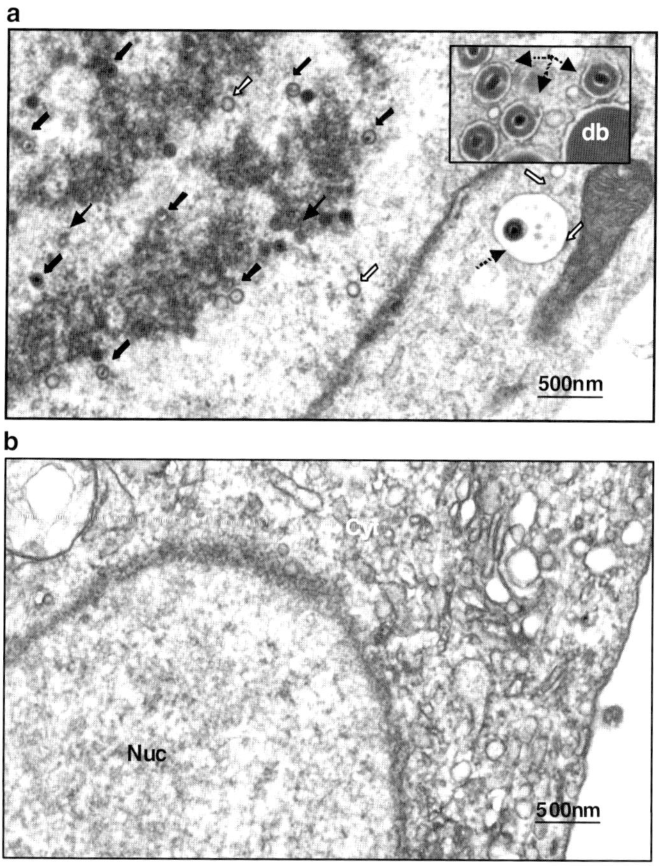

Fig. 4.7 Electron microscopy analysis of HCMV infected human fibroblasts. Infected cells were cultured in the absence (**a**) or presence of 0.5 μM MG132 (**b**) for 72 h prior to preparation for electron microscopy. (**a**) shows areas of the cell with a box from the cytoplasm, (**b**) represents areas of the nucleus and the cytoplasm. White arrows represent A capsids, black arrows B capsids, bold black arrows C capsids, dashed arrows virions. db indicates dense bodies. Scale bars correspond to 500 nm

In order to investigate whether this complete breakdown in virus maturation was caused by suppression of viral protein expression immunofluorescence and Western blot analyses using MG132 were performed. We used antibodies against the immediate early proteins IE1/2, the early proteins pUL44 (the processivity factor of the viral DNA polymerase) and pUL112–113 (protein complex required for formation of pre-replication centers) and the late proteins MCP (major capsid protein) and pp28 (the tegument protein). It was demonstrated that IE gene expression was suppressed at low but not high MOI. Since replication centers no longer exist in MG132 treated cells E gene expression was not only reduced but in addition, the proteins pUL112–113 were translocated to pre-replication compartments (small punctuate structures in Fig. 4.8). In contrast, expression of late structural proteins including the MCP and the true late protein pp28 (Fig. 4.9) was completely abolished at both high and low MOI, thus explaining the absence of progeny virus observed using electron microscopy (Fig. 4.7).

During HCMV replication concatemeric viral DNA is cleaved into unit-length genomes by the HCMV terminase (see above) for packaging into capsids. By using pulsed-field gel electrophoresis (PFGE) of infected cells in the presence of increasing amounts of MG132 it was analyzed whether MG132 also interferes with DNA cleavage. Interestingly the PI MG132 also inhibits cleavage of HCMV DNA in a concentration-dependent manner (Fig. 4.10), thus demonstrating that MG132 interferes with expression of viral E and L genes and prevents the cleavage of concatenated viral DNA, the key step in viral replication.

Our results led us to the question whether any replication occurs in the presence of MG132. Therefore we performed incorporation analyses using the nucleoside bromdesoxyuridin (BrdU). Due to the virus-host shutoff in HCMV infected cells BrdU is preferably incorporated into newly synthesized viral DNA. Interestingly MG132 strongly inhibits *de novo* viral DNA synthesis (Kaspari et al., 2008). In conclusion this proteasome inhibitor interferes with viral replication at a very early stage in the viral life-cycle.

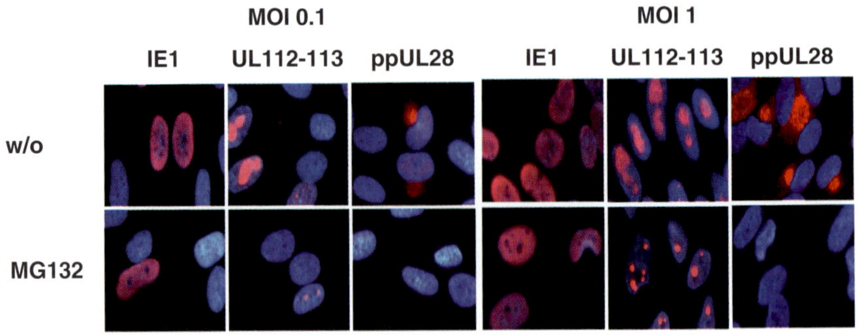

Fig. 4.8 Influence of MG132 on expression of HCMV IE, E and L proteins. HCMV infected fibroblasts (MOI 0.1 or 1) were cultured in the presence (MG132) or absence (w/o) of 0.5 μM MG132 for 72 h and subjected to immunofluorescence using antibodies against IE1, pUL112–113 and ppUL28

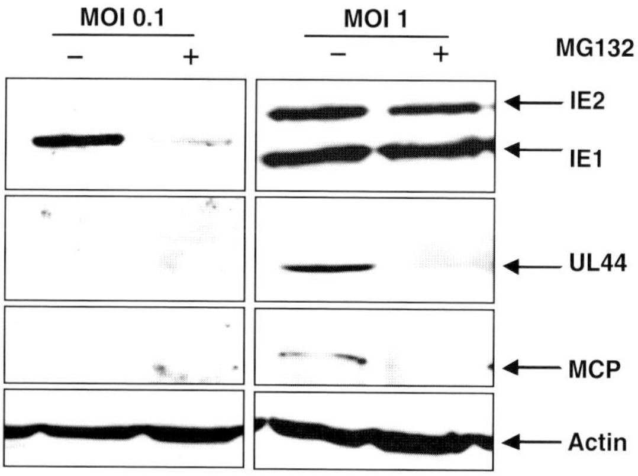

Fig. 4.9 HCMV protein expression in MG132 treated cells is MOI-dependent. HELF were mock- or HCMV infected (MOI 0.1 and 1.0) and cultured with or without 0.5 μM MG132. At 72 h p.i. cell extracts were subjected to Western blot analysis using antibodies against IE1/IE2, pUL44 and MCP. Membranes were re-probed with an antibody against β-actin to verify equal loading. Arrows indicate the position of proteins

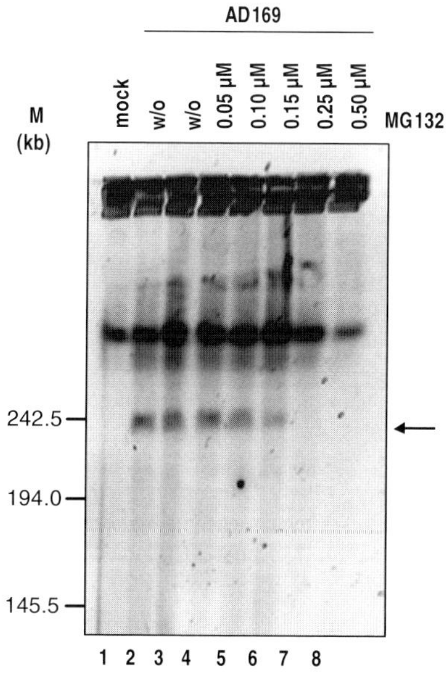

Fig. 4.10 Analysis of viral DNA cleavage. Human fibroblasts were mock- or HCMV infected and cultured in the absence (w/o) or presence of 0.05, 0.10, 0.15, 0.25 and 0.50 μM MG132. At 72 h p.i. cells were harvested for PFGE. The size of HCMV genome monomers (arrow) was determined to be 230 kb using the Lambda Ladder PFG Marker

4.4 Conclusions

Our results indicate that the functionality of the proteasome is essentially required for HCMV replication, thus identifying a new target for HCMV therapy. However, further analyses are required for identification of both the key protein whose stabilization by proteasome inhibitors blocks viral replication and the corresponding E3 ubiquitin ligase. By inhibiting the particular E3 enzyme, interference with HCMV replication would be more specific and unwanted side effects could be avoided.

Acknowledgments We would like to thank all members of the EB lab especially Angelika Lander and Rita Schilf for excellent technical assistance. The research has been supported in part by the Deutsche Forschungsgemeinschaft (Bo 1214/5-4) and the Sonnenfeld Stiftung.

References

Becker, A., Murialdo, H. 1990, Bacteriophage lambda DNA: the beginning of the end. *J Bacteriol.* **172**: 2819–2824.

Bhattacharyya, S.P., Rao, V.B. 1993, A novel terminase activity associated with the DNA-packaging protein gp17 of bacteriophage T4. *Virology*. **196**: 34–44.

Bhattacharyya, S.P., Rao, V.B. 1994, Structural analysis of DNA cleaved in vivo by bacteriophage T4 terminase. *Gene*. **146**: 67–72.

Black, L.W. (1988) DNA packaging in dsDNA bacteriophages. In Calendar, R. (ed.), The Bacteriophages, Plenum, New York.

Bogner, E. 2002, Human cytomegalovirus terminase as a target for antiviral chemotherapy. *Rev Med Virol*. **12**: 115–127.

Bogner, E., Reschke, M., Reis, B., Mockenhaupt, T., Radsak, K. 1993, Identification of the gene product encoded by ORF UL56 of the human cytomegalovirus genome. *Virology*. **196**: 290–293.

Bogner, E., Radsak, K., Stinski, M.F. 1998, The gene product of human cytomegalovirus open reading frame UL56 binds the pac motif and has specific nuclease activity. *J Virol*. **72**: 2259–2264.

Bresnahan, W.A., Boldogh, I., Thompson, A.E., Albrecht, T. 1996, Human Cytomegalovirus inhibits cellular DNA synthesis and arrests productively infected cells in late G1. *Virology*. **224**: 150–160.

Catalano, C.E., Cue, D., Feiss, M. 1995, Virus DNA packaging: the strategy used by phage lambda. *Mol Microbiol*. **16**: 1075–1086.

Ciechanover, A. 1994, The ubiquitin-proteasome proteolytic pathway. *Cell*. **79**: 13–21.

Ciechanover, A. 1998, The ubiquitin-proteasome pathway: the complexity and myriad functions of protein death. *Proc Natl Acad Sci USA*. **95**: 2727–2730.

Coux, O., Tanaka, K., Goldbert, A.L. 1996, Structure and functions of the 20S and 26S proteasomes. *Ann Rev Biochem*. **65**: 2465–2171.

Dick, L.R., Cruikshank, A.A., Destree, A.T., Grenier, L., McCormack, T.A., Melandri, F.D., Nunes, S.L., Palombella, V.J., Parent, L.A., Plamondon, L., Stein, R.L. 1997, Mechanistic studies on the inactivation of the proteasome by lactacystin in cultured cells. *J Biol Chem*. **271**: 7273–7276.

Dittmer, D., Mocarski, E.S. 1997, Human cytomegalovirus infection inhibits G1/S transition. *J Virol*. **71**: 1629–1634.

Feiss, M. 1986, Structure, function and specificity of the DNA packaging signals in double-stranded DNA viruses. *Trends Genet*. **2**: 100–104.

Feiss, M., Becker, A. 1983, DNA packaging and cutting. In Hendrix, R.W., Roberts, J.W., Stahl, F.W., Weisberg, R.A. (eds.), Lambda II, Cold Spring Harbor, New York, pp. 305–330.

Fenteany, G., Standaert, R.F., Lane, W.S., Coi, S., Corey, E.J., Schreiber, S.L. 1995, Inhibition of proteasome activities and subunit-specific amino-terminal threonine modification by lactacystin. *Science*. **268**: 726–731.

Fujisawa, H., Hearing, P. 1994, Structure, function and specificty of the DNA packaging signals in double-stranded DNA viruses. *Seminars Virol*. **5**: 5–13.

Gibson, W. 1996, Structure and assembly of the virion. *Intervirology*. **39**: 389–400.

Guo, P., Peterson, C., Anderson, D. 1987, Prohead and DNA-gp3-dependent ATPase activity of the DNA packaging protein gp16 of bacteriophage phi 29. *J Mol Biol*. **197**: 229–236.

Hang, J.Q., Tack, B.F., Feiss, M. 2000, ATPase center of bacteriophage lambda terminase involved in post-cleavage stages of DNA packaging: identification of ATP-interactive amino acids. *J Mol Biol*. **302**: 777–795.

Harty, R.N., Brown, M.E., McGettigan, J.P., Wang, G., Jayakar, H.R., Huibregtse, J.M., Whitt, M.A., Schnell, M.J. 2001, Rhabdoviruses and the cellular ubiquitin-Proteasome system: a budding interaction. *J Virol*. **75**, 10623–10629.

Hengel, H., Koopmann, J.O., Flohr, T., Muranyi, W., Goulmy, E., Hämmerling, G.J., Koszinowski, U.K., Momburg, F. 1997, A viral ER-resident glycoprotein inactivates the MHC-encoded peptide transporter. *Immunity*. **6**: 623–632.

Hewitt, E.W., Sen Gupta, S., Lehner, P. 2001, The human cytomegalovirus gene product US6 inhibits ATP binding by TAP. *EMBO J*. **20**: 387–396.

Hwang, J.-S., Bogner, E. 2002, ATPase activity of the terminase subunit pUl56 of human cytomegalovirus. *J Biol Chem*. **277**: 6943–6948.

Kaspari, M., Tavalai, N., Stamminger, T., Zimmermann, A., Schilf, R., Bogner, E. 2008, Proteasome inhibitor MG132 blocks viral replication and assembly of human cytomegalovirus. *FEBS Lett*. **582**: 666–672.

Khor, R., McElroy, L.J., Whittaker, G.R. 2003, The ubiquitin-vacuolar protein sorting system is selectively required during entry of influenza virus into host cells. *Traffic*. **4**: 857–868.

La Frazia, S., Mici, C., Santoro, M.G. 2006, Antiviral activity of proteasome inhibitors in herpes simplex virus-1 infection: role of nuclear factor-kB. *Antiviral Ther*. **11**: 995–1004.

Lee, D.H., Goldberg, A.L. 1998, Proteasome inhibitors: valuable new tools for cell biologists. *Trends Cell Biol*. **8**: 397–403.

Mettenleiter, T.C. 2002, Herpesvirus assembly and egress. J. *Virol*. **76**: 1537–1547.

Mocarski, E.S., Courcelle, C.T. 2001, Cytomegalovirus and their replication. In Knipe, D., Howley, T. (eds.), Fields Virology. Lippincott, Williams & Wilkins, Philadelphia, PA, pp. 2629–2673.

Momburg, F., Hengel, H. 2001, Corking the bottle neck: the transporter associated with antigen processing (TAP) as a target for the immune subversion by viruses. *Curr Top Microbiol Immunol*. **269**: 57–74.

Morita, M., Taska, M., Fujisawa, H. 1993, DNA packaging ATPase of bacteriophage T3. *Virology*. **193**: 748–752.

Penfold, M.E., Mocarski, E.S. 1997, Formation of cytomegalovirus DNA replication compartments defined by localization of viral proteins and DNA synthesis. *Virology*. **239**: 46–61.

Radsak, K., Eickmann, M., Mockenhaupt, T., Bogner, E., Kern, H., Eis-Hübinger, A., Reschke, M. 1996, Retrieval of human cytomegalovirus glycoprotein B from infected cell surface for virus envelopment. *Arch Virol*. **141**: 557–572.

Rao, V.B., Black, L.W. 1988, Cloning, overexpression and purification of the terminase proteins gp16 and gp17 of bacteriophage T4. Construction of a defined in-vitro DNA packaging system using purified terminase proteins. *J Mol Biol*. **200**: 475–488.

Roizman, B., Knipe, D. 2001, Herpes simplex viruses and their replication. In Knipe, D.M., Howley, P.M. (eds.), Fields Virology, 3rd ed. Lippincott, Williams & Wilkins, Philadelphia, PA, pp. 2399–2460.

Scheffczik, H., Savva, C.G.W., Holzenburg, A., Kolesnikova, L., Bogner, E. 2002, The terminase subunits pUL56 and pUL89 of human cytomegalovirus are DNA metabolizing proteins with toroidal structure. *Nucl Acids Res*. **30**: 1695–1703.

Scholz, B., Rechter, S., Drach, J.C., Townsend, L.B., Bogner, E. 2003, Identification of the ATP-binding site in the terminase subunit pUL56 of human cytomegalovirus. *Nucl Acids Res*. **31**: 1426–1433.

Schubert, U., Ott, D.E., Chertova, E.N., Welker, R., Tessmer, U., Princiotta, M.F., Bennink, J.R., Kräusslich, H.G., Yewdell, J.W. 2000, Proteasome inhibition interferes with gag polyprotein processing, release, and maturation of HIV-1 and HIV-2. *Proc Natl Acad Sci USA*. **97**: 13057–13062.

Thoma, C., Borst, E., Messerle, M., Rieger, M., Hwang, J.-S., Bogner, E. 2006, Identification of the interaction domain of the small terminase subunit pUL89 with the large subunit pUL56 of human cytomegalovirus. *Biochemistry*. **45**: 8855–8863.

Tomoda, H., Omura, S. 2000, Lactacystin, a proteasome inhibitor: discovery and its application in cell biology. *Yakugaku Zasshi*. **120**: 935–949.

Watanabe, H., Tanaka, Y., Shimazu, Y., Sugahara, F., Kuwayama, M., Hiramatsu, A., Kiyotani, K., Yoshida, T., Sakaguchi, T. 2005, Cell-Specific inhibition of paramyxovirus maturation by proteasome inhibitors. *Microbiol Immunol*. **49**: 835–844.

Wiertz, E.J., Jones, T.R., Sun, L., Bogyoto, M., Geuze, H.J., Ploegh, H.L. 1996, The human cytomegalovirus US11 gene product dislocates MHC class I heavy chains from the endoplasmic reticulum to the cytosol. *Cell*. **84**: 769–779.

Chapter 5
Rational Drug Design of HTLV-I Protease Inhibitors

Jeffrey-Tri Nguyen and Yoshiaki Kiso

Abstract In 2005, it was estimated that 20–30 million people in endemic regions of the world are infected with the human T-cell lymphotrophic virus type-1 (HTLV-I). The virus spreads via breast milk, sexual fluids and blood, to cause adult T-cell lymphocytic leukemia, which progresses to HTLV-associated myelopathy/tropical spastic paraparesis, along with other inflammatory diseases and opportunistic infections. There is no cure and patients die. Although HTLV-I is distantly related to the human immunodeficiency virus (HIV), treating patients with HIV protease inhibitors has been ineffective. In this chapter, we describe our rational drug design method starting with a substrate sequence that is processed by the HTLV-I protease, followed by our initial octapeptidic HTLV-I protease inhibitors containing mostly natural amino acids, from which we derived our small and very potent tetrapeptidic non-natural amino acid inhibitors of HTLV-I protease.

Keywords human T-cell lymphotrophic virus • adult T-cell leukemia • human immunodeficiency virus • aspartic protease inhibitor

5.1 Introduction

Human T-cell lymphotrophic virus type 1 (HTLV-I), also called human T-cell leukemia virus, as its names suggest, is the etiologic agent for T-cell leukemia and T-cell lymphoma. In 1977, Takatsuki and colleagues from Kyoto, Japan first described a strain of oncogenic HTLV-I that causes an aggressive malignancy of primarily $CD4^+$ T-lymphocytes, adult T-cell leukemia/lymphoma (ATL), which presents as skin lesions, lymphadenopathy and hepatosplenomegaly (swollen liver and spleen) (Takatsuki, 2005). However, because ATL is expressed in 2–6%

J.-T. Nguyen and Y. Kiso (✉)
Department of Medicinal Chemistry, Center for Frontier Research in Medicinal Science and 21st Century COE Program, Kyoto Pharmaceutical University, Yamashina-ku, Kyoto 607–8412, Japan
e-mail: kiso@mb.kyoto-phu.ac.jp

U. Lendeckel and N.M. Hooper (eds.), *Viral Proteases and Antiviral Protease Inhibitor Therapy*, Proteases in Biology and Disease 8,

of infected individuals after a latent period of 20–50 years, its concerns have been dismissed by many individuals (Nitta et al., 2006). In the later stage of the disease, HTLV-I infection may worsen to HTLV-associated myelopathy/tropical spastic paraparesis (HAM TSP) which is predominantly associated with $CD8^+$ T-lymphocytes (Jacobson, 2002). HAM/TSP affects the spinal cord at the thoracic level causing a gradual onset of symmetrical weakness and paralysis of the lower limbs, along with upper neuronal signs of dysfunction (Ding et al., 1998). As a result of skin and brain infiltration, several other inflammatory diseases, such as alveolitis (lung), rheumatoid arthritis (joint), infective dermatitis (skin), myositis (muscle), uveitis (eye), and opportunistic infections from *Staphylococcus aureus* and *Strongyloides stercoralis* may also be present (Semmes, 2006).

HTLV-I infects 20–30 million people worldwide (Nicot, 2005). Areas of the world where the virus is endemic include the equatorial regions of Africa, Central and South America, the Middle East, the Caribbean Basin, the Pacific region of Melanesia, and Japan (Proietti et al., 2005). All routes of transmission require close contact with infected T-lymphocytes via breast-feeding (11–40% risk) (Ureta-Vidal et al., 1999), sexual intercourse, and intravenous exposure to blood which includes blood transfusion and the sharing of contaminated injection devices (40–60% risk) (Manns et al., 1992). The virus often spreads to the United States and Europe through injection of illicit drugs and sexual transmission. In Japan, it is estimated that 1.2 million people are infected with HTLV-I with more than 700 newly diagnosed cases each year (Yamaguchi, 1994). Moreover, in Japan, where the main route of transmission is breast milk, 15–25% individuals in high risk groups are viral carriers out of which 6% will die from the infection (Arisawa et al., 2006; Bittencourt, 1998). Moreover, at the present time (2008), unlike HIV and other viral oncogenes, there is no effective treatment to eradicate HTLV-I. The prognosis for ATL patients is poor, with a median survival time of 13 months in aggressive cases (Yamada et al., 2001). In ATL patients, HTLV-infected T-lymphocytes have an innate resistance to apoptosis, and as a result, also to conventional chemotherapy regimens (Ravandi et al., 2005). Only comfort care is available to alleviate the symptoms of the disease. Infection with the virus is life-long.

In 1977, a set of symptoms and infections associated with HTLV-I was discovered by Takatsuki and co-workers (Takatsuki, 2005). A few years later in 1981, the first identified human retrovirus, HTLV-I was isolated in the laboratory of Gallo (Poiesz et al., 1981). Much like its better known relative, the human immunodeficiency virus (HIV), HTLV-I is a human single-stranded ribonucleic acid (RNA) retrovirus (Coffin et al., 1999). A close relative is the bovine leukemia virus (BLV) (Sperka et al., 2007). Interestingly, although the end result of HTLV-I infection is an immunosuppressive effect, HTLV-I, unlike HIV, has an immunostimulating effect (Nishiura et al., 2004). HTLV-I activates Th1 T-helper cells to overproduce Th1 cytokines. Feedback mechanisms to these Th1 cytokines lead to a suppression of the Th2 T-helper cells and Th2 cytokines, resulting in a reduction in the ability of the infected host to mount an adequate Th2-mediated immune response, and thereby causing HAM/TSP and other HTLV-associated diseases. As for the cause of leukemia, we believe that the pro-oncogenic incorporation of viral RNA into human host DNA, along with chronic stimulation to overproduce cytokines may play a role in the development of ATL.

5.2 The HTLV-I Protease

It was not until 1989 that HTLV-I protease, a viral aspartic proteolytic enzyme that plays a critical role in HTLV-I replication, was identified and isolated (Hatanaka and Nam, 1989). Physically, HTLV-I protease resembles HIV protease, in that it is a small homodimer composed of two identical peptide chains that form into a pincer shape where the tip of the pincer is referred to as the flap region and the central cavity as the active site region (Fig. 5.1) (Kadas et al., 2004). A protein is introduced within the active site region, where its cleavage is coordinated by a water molecule and two aspartic acid residues found at the base of the active site of the protease. However, HTLV-I protease is considerably larger (125 amino acids per chain) than HIV protease (98 amino acid per chain), and only shares a 28% sequence identity with HIV protease.

HTLV-I protease is encoded within the viral RNA (Fig. 5.1) (Heidecker et al., 2002). In common with other human retroviruses, the genome of HTLV-I's RNA consists of two flanking long terminal repeats and reading frames in the middle. These reading frames encode for a gag (group antigen, 48 kDa), pro (protease, a 125 amino acid peptide), pol (polymerase, 99 kDa), env (envelope, 54 kDa) and short regulatory genes (Rex, Tax, etc.). Comparatively, both HIV and HTLV-I genes share a similar order, but, considering that the HIV genome is longer than that of HTLV-I, the genes are located at different positions within their respective genome (Ding et al., 1998). In the HTLV-I genome, the genes are overlapped at three ranges, where as the HIV genome only has one overlap with the pro gene being a part of the pol gene. These overlaps are more commonly referred to as "reading frame shifts". In the life cycle of the virus, the genes are transcribed and translated to precursor polyproteins. HTLV-I protease cleaves the Gag precursor polyprotein into matrix (MA), capsid (CA) and nucleocapsid (NC) proteins; another HTLV-I protease (PR) from the peptide sequence; and the Pol precursor polyprotein into reverse transcriptase-ribonuclease H (RT-RH) and integrase (IN) (Fig. 5.1). The third polyprotein, Env, contains two envelope proteins, surface glycoprotein (SU) and transmembrane protein (TM) that are associated with the host cell-derived viral membrane. These resulting proteins are then assembled, developed and matured into a virion that leaves the host cell as a newly formed virus.

The endeavour of our research group is to inhibit HTLV-I protease and ultimately stop viral replication so as to possibly cure HTLV-associated diseases.

As a first step in our research, the HTLV-I protease itself must be produced in the laboratory with sufficient purity and stability such that inhibitory assays could be performed. In 2004, after several failed attempts at reproducing literature procedures to obtain HTLV-I proteases, we designed our own mean of obtaining HTLV-I protease from recombinant and synthetic sources (Maegawa et al., 2004). However, one major drawback with the produced HTLV-I protease is its slow reactivity with our substrate. A substrate is a small peptide that can be digested by the protease. Several research groups have designed and synthesized various substrates based on the sequences of HTLV-I polypeptides (Shuker et al., 2003; Tözsér et al., 2000;

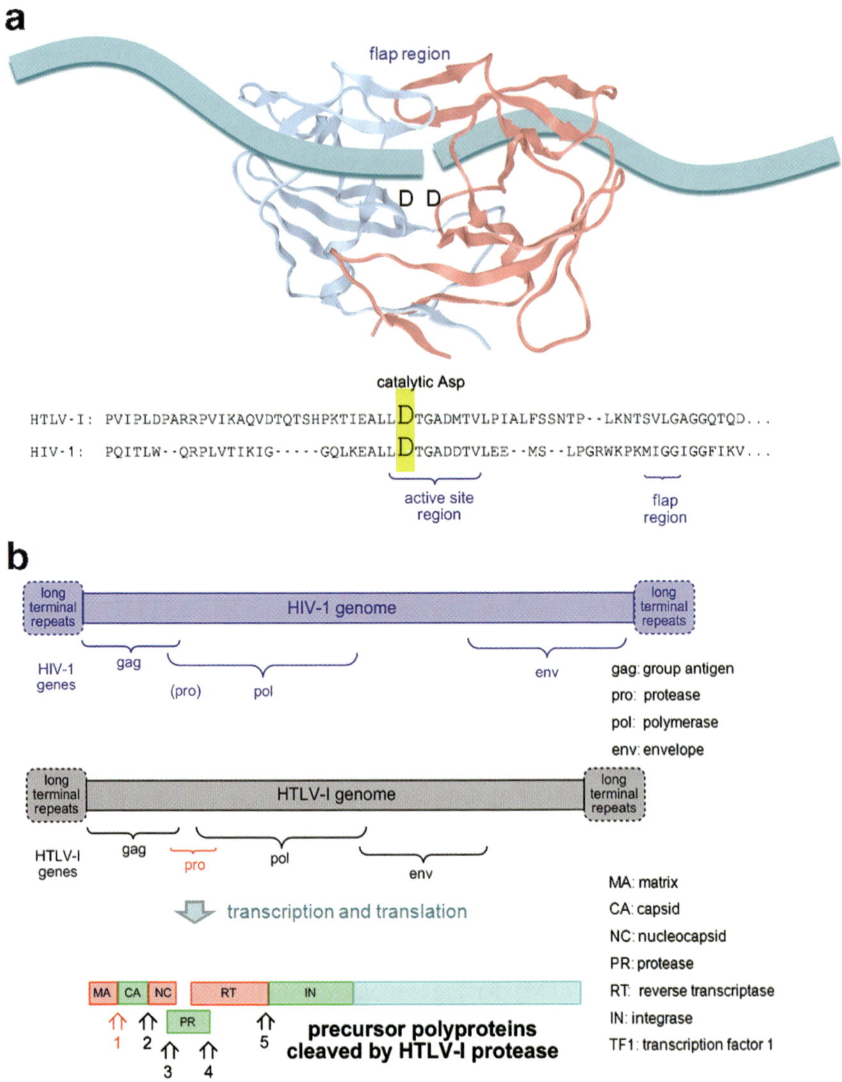

Site	Name	Polyprotein Sequence
1	MA/CA	...-Ala-Pro-Gln-Val-Leu / Pro-Val-Met-His-Pro-...
2	CA/NC	...-Lys-Thr-Lys-Val-Leu / Val-Val-Gln-Pro-Lys-...
3	TF1/PR	...-Asp-Pro-Ala-Ser-Ile-Leu / Pro-Val-Ile-Pro-...
4	PR/p3	...-Lys-Gly-Pro-Pro-Val-Ile-Leu / Pro-Ile-Gln-Ala-Pro-...
5	RT/IN	...-Val-Leu-Gln-Leu / Ser-Pro-Ala-Asp-...

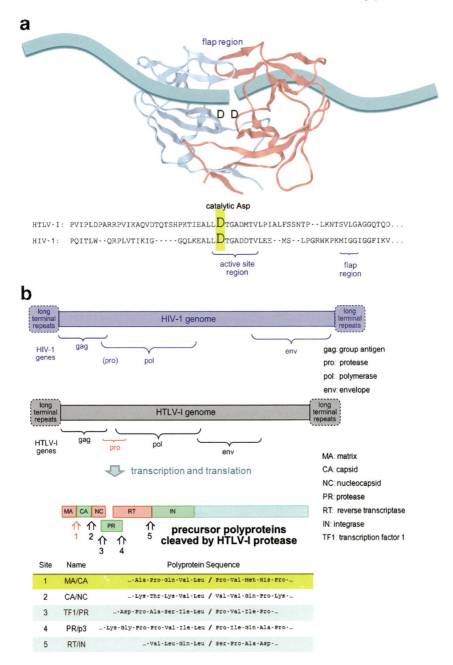

Fig. 5.1 The HTLV-I protease: (**a**) active site and flap regions of a dimeric aspartyl protease, HTLV-I protease; (**b**) HTLV-I genome and protease cleavage sites

Naka et al., 2006; Bang et al., 2007). In 2005, in an effort to obtain reliable X-ray diffraction crystallography data of an HTLV-I protease inhibitor in complex with HTLV-I protease, Wlodawer research group noted that an HTLV-I protease mutation of Leu to Ile at residue 40 would prevent autolysis (Li et al., 2005). Following their lead, in 2008, we developed a more proteolytic HTLV-I protease L40I mutant to increase the reliability of our HTLV-I protease assay. As a result, compared to our former method, incubation time was dramatically reduced from 6 h to 30 min, thus permitting a cost-effective mean of determining IC_{50} values (Nguyen et al., 2008). For that reason, HTLV-I protease inhibitory activity will be presented as a percent inhibition at 100 μM of the test compound and/or IC_{50} values. The laboratories of Aimoto and Akaji also reported a chemically synthesized, autodigestion-resistant HTLV-I protease C2A mutant that was used in their study on HTLV-I protease substrates and inhibitors (Naka et al., 2006; Teruya et al., 2002).

5.3 HIV Protease Inhibitors

As a possible source for lead compounds in HTLV-I protease inhibitor design, we examined HTLV-I's more extensively studied relative, HIV (Louis et al., 1999). Our research group had great successes developing HIV protease inhibitors, and considering that HIV and HTLV-I proteases share a 45% similar sequence identity at the active site region (Fig. 5.2) (Tözsér et al., 2000) several of our super-potent HIV protease inhibitors (88–98% HIV protease inhibition at 50 nM of the test compound) and ritonavir, a potent commercially available HIV protease inhibitor, were examined for inhibitory activity against HTLV-I protease (Maegawa et al., 2004). Unfortunately, the most potent HTLV-I protease inhibitor at 100 μM exhibited only 43% activity against the enzyme. Indeed, Tözsér research group also reported on the ineffectiveness of several other commercial HIV protease inhibitors, namely amprenavir, saquinavir, indinavir and nelfinavir, against HTLV-I protease (Bagossi et al., 2004). Several studies have reported that both substrate specificity and inhibition profile of the two enzymes are fairly different (Ding et al., 1998; Shuker et al., 2003; Louis et al., 1999; Hruskova-Heidingsfeldova et al., 1997). On the other hand, Yamamoto's research group suggested the contrary in that ritanovir was shown to be effective against ATL both in *in vitro* and *in vivo* mouse models, although the group did not directly test for HTLV-I protease inhibitory activity (Dewan et al., 2006). We concluded, with support from X-ray diffraction crystallographic data (Li et al., 2005) that HTLV-I protease inhibitor design cannot be solely based on HIV protease inhibitors.

To inspect the relationship between HIV and HTLV-I protease inhibition by a common compound, we collected all of our comparable inhibitors and plotted a chart to compared HIV and HTLV-I protease inhibition (Fig. 5.2). From the plotted 128 compounds, we observed that, in general, HTLV-I protease inhibitors also exhibit potent HIV protease inhibitory activity. However, compounds with potent HIV protease inhibitory activity do not necessarily exhibit potent HTLV-I protease

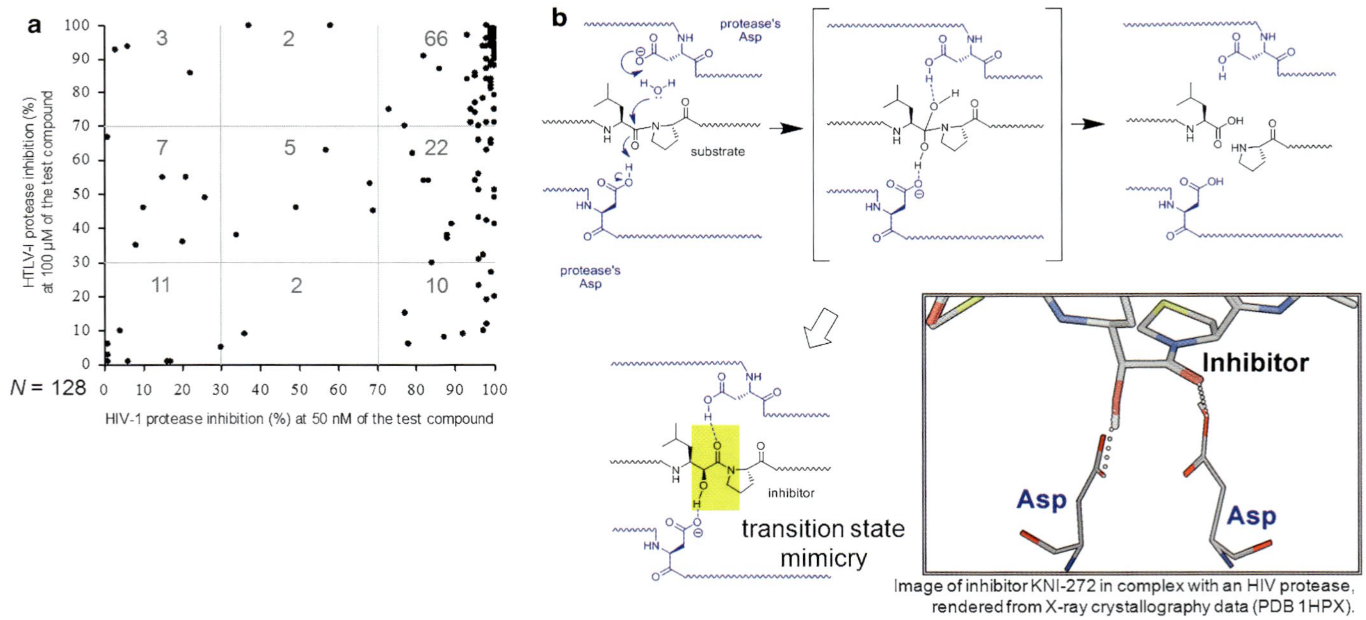

Fig. 5.2 (**a**) The relationship between HTLV-I and HIV protease inhibition. (**b**) Transition state mimicry of substrate hydrolysis

inhibitory activity. Our results reflect similar observations made by Tözsér and co-workers that HTLV-I protease displays a high degree of specificity over that of HIV protease (Kadas et al., 2004). Whereas Tözsér research group reported narrower activation profiles from substrates, we observed narrower inhibition profiles from our HTLV-I protease inhibitors. Hence, differences in the amino acid sequences of HIV and HTLV-I proteases produce differences in susceptibilities to inhibitors, in that HIV protease is more susceptible to different substrates and inhibitors than HTLV-I protease. On a side note, our past studies have shown that HIV protease is also less specific than plasmepsin II, an aspartic protease associated with malaria (Nezami et al., 2002, 2003; Kiso et al., 2004; Abdel-Rahman et al., 2004; Hidaka et al., 2007). These observations imply that should the need arise, HTLV-I protease inhibitors and plasmepsin inhibitors could become lead compounds in the design of novel HIV protease inhibitors.

5.4 The Central Inhibitory Unit

The designs of our initial HTLV-I protease inhibitors were based on the sequences of HTLV-I polyproteins (Fig. 5.3). HTLV-I protease cleaves polyproteins at different sites (Heidecker et al., 2002). Bearing in mind that the design of our HTLV-I protease inhibitors are based on substrate sequences, information about readily cleavable substrates are relevant and invaluable in the pursuit of HTLV-I protease inhibitors. Of the five known sites, three sites, MA/CA, TF1(transcription factor 1)/PR and PR/p3(tripeptide), have Leu-Pro at the point of cleavage (Shuker et al., 2003). Considering that Leu-Pro cleavage is viral specific, we arbitrarily chose a MA/CA substrate as a reference to design our inhibitor. Our peptidic inhibitors would compete for the interactions of the substrate and protease.

The central inhibitory unit of our HTLV-I protease inhibitors is based on the mimicry of a transition state formed during amide hydrolysis by the protease (Fig. 5.2). The most widely accepted mechanism is an acid-based system involving two active aspartic acid residues in the active site, a water molecule that resides between the substrate and inhibitor, and a Leu-Pro containing substrate. These two aspartic acid residues act respectively as a proton donor and acceptor to catalyze the hydrolysis of peptide bonds in the substrate. The water molecule is partly activated by an aspartate and makes a nucleophilic attack at the Leu carbonyl carbon in the substrate. The Leu carbonyl oxygen then captures a proton from the other aspartic acid residue in the active site, therefore forming a tetrahedral intermediate. This important intermediate is the transition state from which our inhibitory unit is designed. Re-stabilizing from the transition state, the amino moiety from the Pro portion becomes a better leaving group, and the substrate is cleaved into two peptide fragments. In designing aspartic protease inhibitors, we envisioned the synthesis of a peptidic substrate analogue that contains a non-hydrolysable transition state isostere in place of the normal hydrolysable Leu P_1-Pro P_1' amide bond.

Fig. 5.3 (**a**) Our octapeptidic HTLV-I protease inhibitors. (**b**) Our hexapeptidic HTLV-I protease inhibitors. (**c**) Akaji research group's tetrapeptidic HTLV-I protease inhibitors. The methods used to determine IC_{50} values were different from ours, Kiso research group, and thus values should not be compared

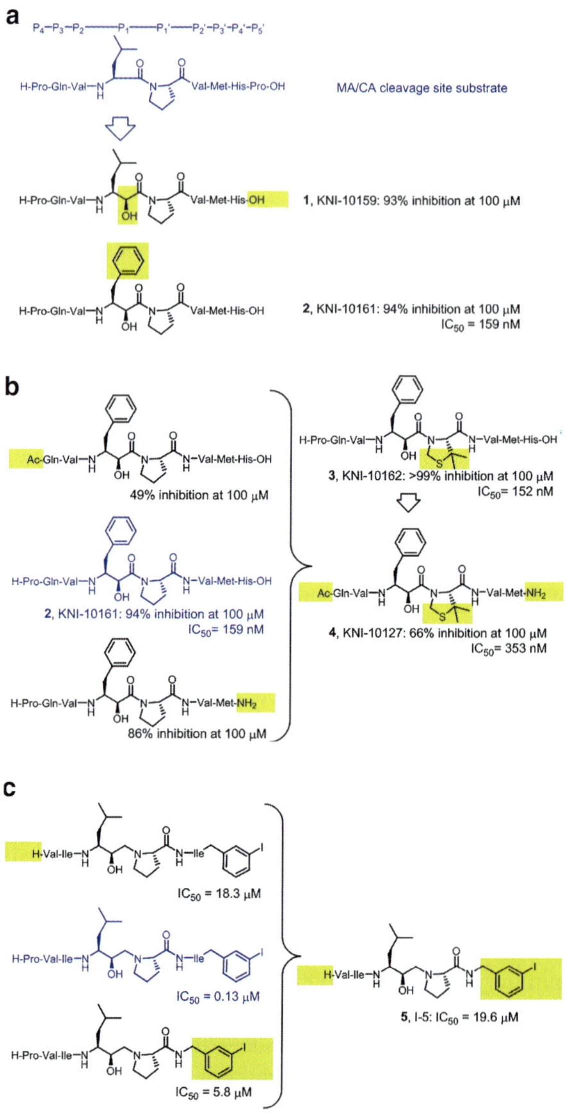

In naming residues and subsites in the active sites, for example, P_2, P_1, P_1' and P_2', the numbering indicates the position of the inhibitor's residue or subsite relative to the catalytic cleavage site, where residues attached to the P_1 C-terminus are referred to as primes. When a hydroxymethylcarbonyl isostere, a functional group that can mimic the tetrahedral transition state formed during amide bond hydrolysis but cannot itself be hydrolysed by the protease, is introduced in a substrate-like compound, the protease's usual biochemical actions do not occur. The critical hydroxyl group, as a transition state mimic, forms hydrogen bond interactions with the carboxylic acid moieties of the two catalytic aspartic acid residues in the protease. The carbonyl oxygen in the inhibitor is also involved with one of the aspartic acids to form a hydrogen bond network. In 1996, for the first time in literature, the existence of this central inhibitory unit and its hydrogen network was proven by our NMR study of an HIV protease-inhibitor complex (Wang et al., 1996a, b; Kiso, 1996; Kiso et al., 1998, 1999). Several X-ray diffraction crystallographic data provided concrete evidence of the hydrogen bond network that is formed between the hydroxymethylcarbonyl isostere inhibitory unit and the protease active site with mediation from a water molecule, as seen with HIV protease (Baldwin et al., 1995; Vega et al., 2004; Doi et al., 2004; Clemente et al., 2006; Lafont et al., 2007) and the malaria proteases, plasmepsins (Abdel-Rahman et al., 2004).

Based on the polyprotein sequence around the MA/CA cleavage site, one of our initial compounds turned out to be a very potent octapeptidic HTLV-I protease inhibitor (**1**), KNI-10159, in which an allonorstatine residue (2*S*,3*S*)-3-amino-2-hydroxy-5-methylhexanoic acid, serves as the hydroxymethylcarbonyl isostere inhibitory unit (Fig. 5.3) (Maegawa et al., 2004). Changing the inhibitory unit to allophenylnorstatine (2*S*,3*S*)-3-amino-2-hydroxy-4-phenyl-butyric acid, afforded an equipotent inhibitor, KNI-10161 (**2**). In view of the larger volume provided by the allophenylnorstatine residue than the allonorstatine residue, we reasoned that the allophenylnorstatine would fill the S_1 subsite more efficiently. Moreover, Tözsér research group, while studying HTLV-I substrates, revealed that a P_1 Phe, which is structurally similar to allophenylnorstatine, is better accommodated by the enzyme's S_1 subsite than a P_1 Leu, which is structurally similar to allonorstatine (Tözsér et al., 2000). Contrarily, in a recent work, Akaji and coworkers' findings suggested that Leu, along with Met and Cys, are more favoured by the S_1 subsite than Phe (Bang et al., 2007). Resuming our discussion on inhibitor KNI-10161 (**2**), changing the P_1 residue from allophenylnorstatine to its diastereomer phenylnorstatine (2*R*,3*S*)-3-amino-2-hydroxy-4-phenylbutyric acid, decreased HTLV-I potency against HTLV-I protease (Nguyen et al., 2008). When the P_1' residue was modified in inhibitor KNI-10162 (**3**), we noted that a more conformationally constrained and bulkier P_1' (*R*)-5, 5-dimethyl-1,3-thiazolidine-4-carboxylic acid residue exhibited slightly more potent inhibitory activity than a P_1' Pro residue (Fig. 5.3). Indeed, the incorporation of a non-natural amino acid P_1' dimethylthiazolidine residue was designed to avoid premature Pro recognition and degradation by other proteases.

5.5 A Truncation Study on HTLV-I Protease Inhibitors

On the design of substrate-based HTLV-I inhibitors, we always try to reduce the size of the inhibitor to potentially improve dissolution, cellular penetration, membrane permeability and intestinal absorption. We performed a truncation study on octapeptidic inhibitor KNI-10161 (**2**) by methodically "removing" the N- and

C-terminal residues to determine the critical points at which inhibitory activity is nearly absent (Fig. 5.3) (Kimura et al., 2007). From the results of the size reduction study and KNI-10162 (**3**)'s P_1' residue discovery, we designed and obtained an expected less potent hexapeptidic HTLV-I protease inhibitor, KNI-10127 (**4**).

Akaji and co-workers also designed and synthesized HTLV-I protease inhibitors, based around the PR/Pol cleavage site, instead of the MA/CA cleavage site (Fig. 5.3) (Naka et al., 2006). The results from their truncation study revealed that their shortest potent HTLV-I protease inhibitor I-5 (**5**) contained four residues with a large 3-iodobenzyl P_1'-capping moiety. It should be noted that the inhibitory activities of our inhibitors and that of Akaji research group should not be compared, because of differences in inhibitory assay methods. Unfortunately, other than Akaji-Aimoto and our Kiso research groups, there has been very few literature reports on novel HTLV-I protease inhibitor development (Tözsér and Weber, 2007).

5.6 Amino Acid Substitution Studies on HTLV-I Protease Inhibitors

HTLV-I protease cleaves at different precursor polyprotein sites and, as mentioned, HTLV-I protease inhibitors were successfully designed around the MA/CA and PR/Pol cleavage sites. Taking this into consideration, the natural amino acid residues of inhibitor KNI-10127 (**4**), namely the P_3-P_2 and P_2'-P_3' residues, were systematically substituted with residues from a small library of natural amino acids (Fig. 5.4) (Kimura et al., 2007). The content of this natural amino acid library was derived from various HTLV-I protease substrates. The inhibitory profiles of our series of compounds suggested that Ile was preferred at multiple subsites in the enzyme. Substrate cleavage efficiency optimization studies by other research groups (Tözsér et al., 2000; Li et al., 2005) supported our findings that Ile was well accommodated at the S_3, S_2 and S_2' subsites (Kimura et al., 2007). Hence, the substitution study with natural amino acids led us to a potent HTLV-I protease inhibitor, KNI-10166 (**6**).

We pursued our substitution study with non-natural amino acid residues, in order to remove peptidic features from the design so as to prevent premature proteolysis by other proteases (Nguyen et al., 2008). In lieu of the Ile residues that were present throughout inhibitor KNI-10166 (**6**), we selected L-*tert*-leucine and L-(+)-α-phenylglycine as possible isosteres of Ile (Fig. 5.4). We methodically substituted each Ile residue in inhibitor KNI-10166 (**6**) to derive potent HTLV-I protease inhibitor KNI-10220 (**7**). At this point in the research, we realized that the P_3' residue had become a lower determinant of potency, because a P_3' replacement of the non-polar Met residue in inhibitor KNI-10220 (**7**) by a polar amino acid, Gln, maintained high inhibitory activity against HTLV-I protease. Moreover, Akaji and colleagues had reported active HTLV-I protease inhibitors lacking a P_3' residue, such as inhibitor I-5 (**5**) (Fig. 5.3). We subsequently removed the P_3' residue to uncover a moderately potent, pentapeptidic HTLV-I protease inhibitor, KNI-10247 (**8**) (Fig. 5.4).

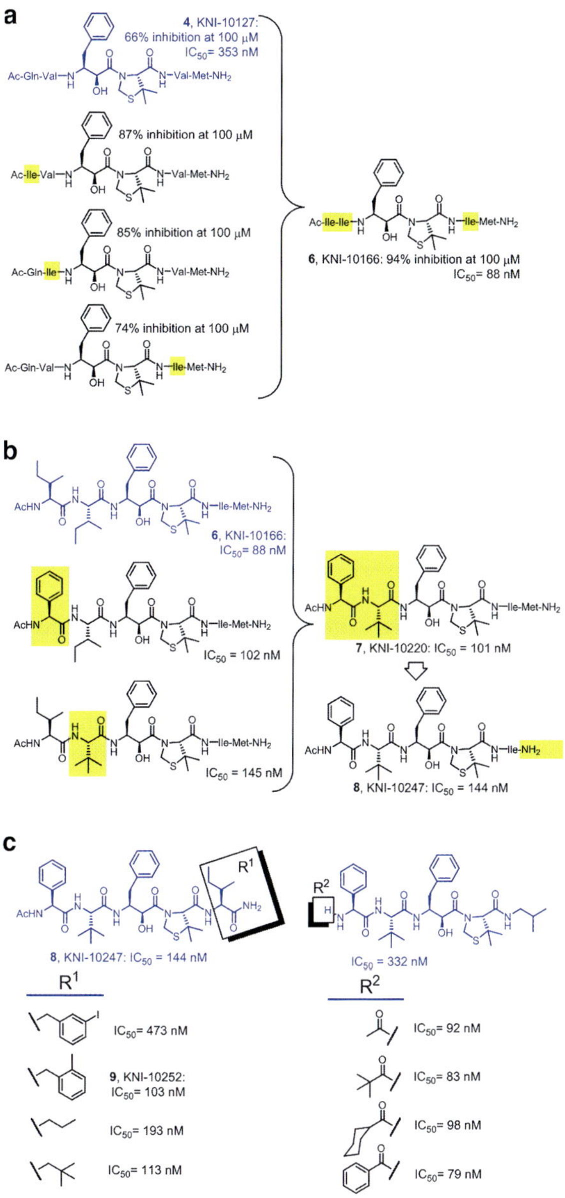

Fig. 5.4 (**a**) Natural amino acid substitution study on hexapeptidic HTLV-I protease inhibitor KNI-10127. (**b**) Non-natural amino acid substitution study on hexapeptidic HTLV-I protease inhibitor KNI-10166. (**c**) Optimization of the P_1'- and P_3-capping groups in HTLV-I protease inhibitors

5.7 P_1'- and P_3-Capping Groups in HTLV-I Protease Inhibitors

The P_2 Ile residue remained as the only natural amino acid present in our HTLV-I protease inhibitor design of compound KNI-10247 (**8**) (Nguyen et al., 2008). We opted to replace the P_2 natural amino acid residue by a P_1'-capping moiety, so as to prevent premature recognition and digestion by the body's endogeneous proteases (Fig. 5.4). In a preliminary study on HTLV-I protease inhibitor, Akaji and co-workers described an inhibitor possessing a 3-iodobenzyl moiety as the P_1'-capping moiety, namely I-5 (**5**) (Fig. 5.3) (Naka et al., 2006). However, when the moiety was applied as a P_1'-cap for inhibitor KNI-10247 (**8**), the resulting compound exhibited low inhibitory potency against HTLV-I protease (Nguyen et al., 2008). Further P_1'-capping moiety optimization using a benzyl ring system led to a highly potent inhibitor KNI-10252 (**9**).

As a different method for optimizing the P_1'-capping moiety, we approached the P_1'-capping moiety from an Ile isostere perspective (Nguyen et al., 2008). We began with an *n*-propyl P_1'-capping moiety and gradually increased the volumetric bulk at the β-carbon, and observed a general trend of enhanced HTLV-I protease inhibitory potency that is proportional with increased bulk. This observation was statistically confirmed by performing a thorough quantitative structure-activity relationship study (Zhang et al., 2008a). Moreover, methyl branching at the P_1'-capping moiety's (*R*)-α-carbon, along with lengthening the P_1'-capping moiety have undesired effects on enzyme inhibition.

As for the P_3-capping moiety, the P_3-cap's carbonyl oxygen is critical in maintaining high inhibitory activity against HTLV-I protease because of important hydrogen bond interactions with the enzyme's Leu57A residue (Zhang et al., 2008a, b). However, the nature of the P_3-cap's hydrophobic component does not contribute as much inhibitory potency as the P_1'-capping moiety. Indeed, any hydrophobic group as a part of the P_3-capping moiety, from a methyl to cyclohexyl or phenyl functional group, would produce highly efficient inhibition against HTLV-I protease.

5.8 Computer-Assisted Docking Simulations of HTLV-I Protease and Its Inhibitors

To date (2008), due to difficulties in crystallizing an inhibitor in complex with HTLV-I protease, only a single complete set of X-ray diffraction crystallographic data was reported (Li et al., 2005). The data described a large decapeptidic inhibitor in complex with a truncated 116 residue L40I mutant protease instead of the full 125 residue HTLV-I protease. Although the crystallized inhibitor was considerably different from our inhibitors, we succeeded in generating several models of our inhibitors in complex with the truncated L40I mutant HTLV-I protease, by also referring to the X-ray diffraction crystallographic coordinates of potent HIV protease inhibitor KNI-577 in complex with its protease (Nguyen et al., 2008; Kimura et al., 2007; Zhang et al., 2008a, b). These models visually clarify the observed trends in inhibitory activities of our inhibitors against HTLV-I protease.

Our more potent HTLV-I protease tetrapeptidic inhibitor KNI-10635 (**10**) was placed inside an HTLV-I protease by computer-assisted docking simulations (Fig. 5.5) (Zhang et al., 2008a). Multiple possible hydrogen bond interactions exist throughout the backbone of the inhibitor and HTLV-I protease's Asp32A,

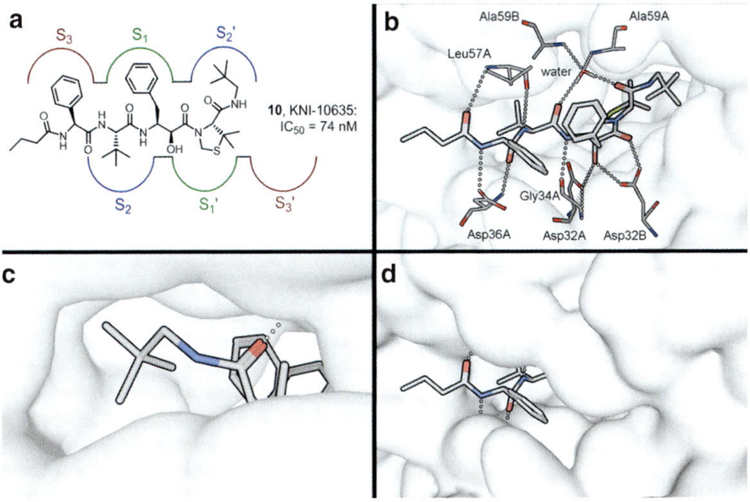

Fig. 5.5 Computer model of KNI-10635 in the active site of HTLV-I protease. Dotted lines represent possible hydrogen bond interactions. (**a**) A cartoon representation of a docked KNI-10635. (**b**) The surface of some HTLV-I protease heavy atoms is hidden to clearly depict various hydrogen bond interactions. (**c**) The P_1'-capping moiety is located deep within the S_2' subsite than (**d**) the P_3-capping moiety that resides near the exposed surface of HTLV-I protease

Asp32B, Gly34A, Asp36A, Leu57A, Ala59A, and Ala59B of each respective dimer. These hydrogen bond interactions anchor the inhibitor in its docked position within the HTLV-I protease. Similar to HIV and plasmepsin protease inhibitors, a water molecule mediates hydrogen bond interactions between the inhibitor and the hairpin regions of HTLV-I protease's flaps via Ala59A and Ala59B. Moreover, the inhibitor's transition state mimic, the hydroxymethylcarbonyl isosteric moiety, effectively interacts with the protease's catalytic Asp32A and Asp32B. As a result, a hydrogen bond network is formed in the vicinity of the critical P_1–P_1' residues.

As reported by Tözsér and colleagues that the S_1 subsite is relatively large and hydrophobic (Tözsér et al., 2000) the phenyl ring of P_1's allophenylnorstatine can be accommodated by the S_1 subsite. As the S_1' subsite appears to be somewhat smaller than the S_1 subsite due to the asymmetry introduced by the peptide bond, the thiazolyl ring of P_1' dimethylthiazoline partially occupies the S_1' subsite with the dimethyl groups pointing toward the shaft of the tunnel. The laboratory of Tözsér also described the S_2 and S_2' subsites as being large and hydrophobic with preference for β-branching on the side-chain. The *tert*-leucine P_2 residue and *tert*-pentyl P_1'-capping moiety fulfill these requirements. The S_3 subsite is larger than the S_2 and S_2' subsites, and is partly exposed to the solvent at the surface of the enzyme. Moreover, a P_3 side-chain can interact with either the more polar surface residues of the protease, or the hydrophobic internal residues within the S_3 subsite. Our docking experiments revealed that the P_3 phenylglycine moiety in KNI-10635, which is relatively larger than the P_2 *tert*-leucine residue, is snugly wedged in a crevasse formed at the S_3 subsite. The S_4 subsite, being more exposed to the solvent at the surface of HTLV-I protease, has less stringent requirements for van der Waal's interactions, hydrophilic and hydrophobic interactions. As such, we observed that the P_3-capping moiety, other than the carbonyl oxygen involved in hydrogen bond interactions with Leu57A, has the least significant effect, relative to the other positions in the inhibitor, on HTLV-I protease inhibitory activity. In simpler words, any modification at the P_3-capping moiety, beyond the carbonyl group, would have the least impact on HTLV-I protease inhibition.

5.9 Conclusion: Outline of HTLV-I Protease Inhibitor Development

Based on the sequence of a substrate that is cleaved by HTLV-I protease, we designed an inhibitor in which the substrate's catalytic residue was replaced by an inhibitory unit (Fig. 5.6). This HTLV-I protease inhibitor was reduced in size to increase the chance for cellular penetration. Each natural amino acid residues in the inhibitor was then optimized to increase inhibitory potency against HTLV-I protease. To prevent premature proteolytic degradation by other proteases, non-natural amino acid residues replaced the natural amino acid units found in the

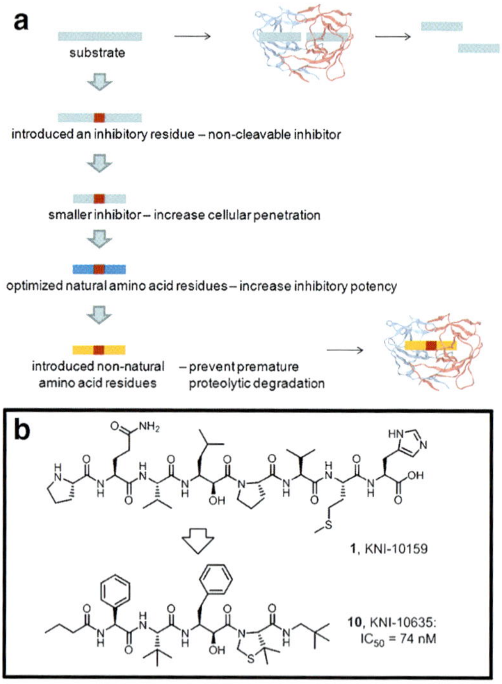

Fig. 5.6 (**a**) Simplified flow-chart of HTLV-I protease inhibitor drug design strategies. (**b**) Optimization of an octapeptidic HTLV-I protease inhibitor, KNI-10159, to a very potent tetrapeptidic compound, KNI-10635

inhibitor. Hence, from our first large, potent octapeptidic HTLV-I protease inhibitor, KNI-10159 (**1**), mostly containing natural amino acid residues, we refined the inhibitor to a small yet very potent tetrapeptidic inhibitor, KNI-10635 (**10**), that is completely composed of non-natural amino acid units.

Acknowledgements The writing of this manuscript was supported in parts by The Frontier Research Program, The 21st Century COE Program from The Ministry of Education, Culture, Sports, Science and Technology, Japan; and the relentless efforts of many researchers and patients to overcome the ravages caused by HTLV-I.

References

Abdel-Rahman, H.M., Kimura, T., Hidaka, K., Kiso, A., Nezami, A., Freire, E., Hayashi, Y., Kiso, Y. 2004, Design of inhibitors against HIV, HTLV-I, and Plasmodium falciparum aspartic proteases. *Biol Chem.* **385**: 1035–1039.

Arisawa, K., Soda, M., Akahoshi, M., Fujiwara, S., Uemura, H., Hiyoshi, M., Takeda, H., Kashino, W., Suyama, A. 2006, Human T-cell lymphotropic virus type-1 infection and risk of cancer: 15.4 year longitudinal study among atomic bomb survivors in Nagasaki, Japan. *Cancer Sci.* **97**: 535–539.

Bagossi, P., Kádas, J., Miklóssy, G., Boross, P., Weber, I.T. and Tözsér, J. 2004, Development of a microtiter plate fluorescent assay for inhibition studies on the HTLV-1 and HIV-1 proteinases. *J Virol Methods.* **119**: 87–93.

Baldwin, E.T., Bhat, T.N., Gulnik, S., Liu, B., Topol, I.A., Kiso, Y., Mimoto, T., Mitsuya, H., Erickson, J.W. 1995, Structure of HIV-1 protease with KNI-272, a tight-binding transition-state analog containing allophenylnorstatine. *Structure.* **3**: 581–590.

Bang, J.K., Teruya, K., Aimoto, S., Konno, H., Nosaka, K., Tatsumi, T., Akaji, K. 2007, Studies on substrate specificity at PR/p3 cleavage site of HTLV-1 protease. *Int J Pept Res Ther.* **13**: 173–179.

Bittencourt, A.L. 1998, Vertical transmission of HTLV-I/II: a review. *Rev. Inst. Med. Trop. Sao Paulo.* **40**: 245–251.

Clemente, J.C., Govindasamy, L., Madabushi, A., Fisher, S.Z., Moose, R.E., Yowell, C.A., Hidaka, K., Kimura, T., Hayashi, Y., Kiso, Y., Agbandje-McKenna, M., Dame, J.B., Dunn, B.M., McKenna, R. 2006, Structure of the aspartic protease plasmepsin 4 from the malarial parasite Plasmodium malariae bound to an allophenylnorstatine-based inhibitor. *Acta Crystallogr., Sect. D.* **62**: 246–252.

Coffin, J.M., Hughes, S.H. and Varmus, H.E. 1999, *Retroviruses.* Woodbury, NY: CSHL, p. 520.

Dewan, M.Z., Uchihara, J.N., Terashima, K., Honda, M., Sata, T., Ito, M., Fujii, N., Uozumi, K., Tsukasaki, K., Tomonaga, M., Kubuki, Y., Okayama, A., Toi, M., Mori, N., Yamamoto, N. 2006, Efficient intervention of growth and infiltration of primary adult T-cell leukemia cells by an HIV protease inhibitor, ritonavir. *Blood.* **107**: 716–724.

Ding, Y.S., Rich, D.H. and Ikeda, R.A. 1998, Substrates and inhibitors of human T-cell leukemia virus type I protease. *Biochemistry.* **37**: 17514–17518.

Doi, M., Kimura, T., Ishida, T., Kiso, Y. 2004, Rigid backbone moiety of KNI-272, a highly selective HIV protease inhibitor: methanol, acetone and dimethylsulfoxide solvated forms of 3-[3-benzyl-2-hydroxy-9-(isoquinolin-5-yloxy)-6-methylsulfanylmethyl-5,8-dioxo-4,7-diazanonanoyl]-*N-tert*-butyl-1,3-thiazolidine-4-carboxamide. *Acta Crystallogr., Sect. B.* **60**: 433–437.

Hatanaka, M. and Nam, S.H. 1989, Identification of HTLV-I gag protease and its sequential processing of the gag gene product. *J Cell Biochem.* **40**: 15–30.

Heidecker, G., Hill, S., Lloyd, P.A., Derse, D. 2002, A novel protease processing site in the transframe protein of human T-cell leukemia virus type 1 PR76(gag-pro) defines the N terminus of RT. *J Virol*. **76**: 13101–13105.

Hidaka, K., Kimura, T., Tsuchiya, Y., Kamiya, M., Ruben, A.J., Freire, E., Hayashi, Y., Kiso, Y. 2007, Additional interaction of allophenylnorstatine-containing tripeptidomimetics with malarial aspartic protease plasmepsin II. *Bioorg Med Chem Lett*. **17**: 3048–3052.

Hruskova-Heidingsfeldová, O., Bláha, I., Urban, J., Strop, P., Pichová, I. 1997, Substrates and inhibitors of human T-cell leukemia virus type 1 (HTLV-1) proteinase. *Leukemia*. **11(Suppl. 3)**: 45–46.

Jacobson, S. 2002, Immunopathogenesis of human T cell lymphotropic virus type I-associated neurologic disease. *J Infect Dis*. **186(Suppl. 2)**: 187–192.

Kadas, J., Weber, I.T., Bagossi, P., Miklóssy, G., Boross, P., Oroszlan, S., Tözsér, J. 2004, Narrow substrate specificity and sensitivity toward ligand-binding site mutations of human T-cell Leukemia virus type 1 protease. *J Biol Chem*. **279**: 27148–27157.

Kimura, T., Nguyen, J.T., Maegawa, H., Nishiyama, K., Arii, Y., Matsui, Y., Hayashi, Y., Kiso, Y. 2007, Chipping at large, potent human T-cell leukemia virus type 1 protease inhibitors to uncover smaller, equipotent inhibitors. *Bioorg Med Chem Lett*. **17**: 3276–3280.

Kiso, A., Hidaka, K., Kimura, T., Hayashi, Y., Nezami, A., Freire, E., Kiso, Y. 2004, Search for substrate-based inhibitors fitting the S_2' space of malarial aspartic protease plasmepsin II. *J Peptide Sci*. **10**: 641–647.

Kiso, Y. 1996, Design and synthesis of substrate-based peptidomimetic human immunodeficiency virus protease inhibitors containing the hydroxymethylcarbonyl isostere. *Biopolymer*. **40**: 235–244.

Kiso, Y., Yamaguchi, S., Matsumoto, H., Mimoto, T., Kato, R., Nojima, S., Takaku, H., Fukazawa, T., Kimura, T., Akaji, K. 1998, KNI-577, a potent small-sized HIV protease inhibitor based on the dipeptide containing the hydroxymethylcarbonyl isostere as an ideal transition-state mimic. *Arch Pharm (Weinheim)*. **331**: 87–89.

Kiso, Y., Matsumoto, H., Mizumoto, S., Kimura, T., Fujiwara, Y., Akaji, K. 1999, Small dipeptide-based HIV protease inhibitors containing the hydroxymethylcarbonyl isostere as an ideal transition-state mimic. *Biopolymer*. **51**: 59–68.

Lafont, V., Armstrong, A.A., Ohtaka, H., Kiso, Y., Mario Amzel, L., Freire, E. 2007, Compensating enthalpic and entropic changes hinder binding affinity optimization. *Chem Biol Drug Des*. **69**: 413–422.

Li, M., Laco, G.S., Jaskolski, M., Rozycki, J., Alexandratos, J., Wlodawer, A., Gustchina, A. 2005, Crystal structure of human T cell leukemia virus protease, a novel target for anticancer drug design. *Proc Natl Acad Sci USA*. **102**: 18332–18337.

Louis, J.M., Oroszlan, S. and Tözsér, J. 1999, Stabilization from autoproteolysis and kinetic characterization of the human T-cell leukemia virus type 1 proteinase. *J Biol Chem*. **274**: 6660–6666.

Maegawa, H., Kimura, T., Arii, Y., Matsui, Y., Kasai, S., Hayashi, Y., Kiso, Y. 2004, Identification of peptidomimetic HTLV-I protease inhibitors containing hydroxymethylcarbonyl (HMC) isostere as the transition-state mimic. *Bioorg Med Chem Lett*. **14**: 5925–5929.

Manns, A., Wilks, R.J., Murphy, E.L., Haynes, G., Figueroa, J.P., Barnett, M., Hanchard, B., Blattner, W.A. 1992, A prospective study of transmission by transfusion of HTLV-I and risk factors associated with seroconversion. *Int J Cancer*. **51**: 886–891.

Naka, H., Teruya, K., Bang, J.K., Aimoto, S., Tatsumi, T., Konno, H., Nosaka, K., Akaji, K. 2006, Evaluations of substrate specificity and inhibition at PR/p3 cleavage site of HTLV-1 protease. *Bioorg Med Chem Lett*. **16**: 3761–3764.

Nezami, A., Luque, I., Kimura, T., Kiso, Y., Freire, E. 2002, Identification and characterization of allophenylnorstatine-based inhibitors of plasmepsin II, an antimalarial target. *Biochemistry*. **41**: 2273–2280.

Nezami, A., Kimura, T., Hidaka, K., Kiso, A., Liu, J., Kiso, Y., Goldberg, D.E., Freire, E. 2003, High-affinity inhibition of a family of Plasmodium falciparum proteases by a designed adaptive inhibitor. *Biochemistry*. **42**: 8459–8464.

Nguyen, J.-T., Zhang, M., Kumada, H.O., Itami, A., Nishiyama, K., Kimura, T., Cheng, M., Hayashi, Y., Kiso, Y. 2008, Truncation and non-natural amino acid substitution studies on HTLV-I protease hexapeptidic inhibitors. *Bioorg Med Chem Lett*. **18**: 366–370.

Nicot, C. 2005, Current views in HTLV-I-associated adult T-cell leukemia/lymphoma. *Am J Hematol*. **78**: 232–239.

Nishiura, Y., Nakamura, T., Fukushima, N., Moriuchi, R., Katamine, S., Eguchi, K. 2004, Increased mRNA expression of Th1-cytokine signaling molecules in patients with HTLV-I-associated myelopathy/tropical spastic paraparesis. *Tohoku J Exp Med*. **204**: 289–298.

Nitta, T., Kanai, M., Sugihara, E., Tanaka, M., Sun, B., Nagasawa, T., Sonoda, S., Saya, H., Miwa, M. 2006, Centrosome amplification in adult T-cell leukemia and human T-cell leukemia virus type 1 Tax-induced human T cells. *Cancer Soc*. **97**: 836–841.

Poiesz, B.J., Ruscetti, F.W., Reitz, M.S., Kalyanaraman, V.S., Gallo, R.C. 1981, Isolation of a new type C retrovirus (HTLV) in primary uncultured cells of a patient with Sézary T-cell leukaemia. *Nature*. **294**: 268–271.

Proietti, F.A., Carneiro-Proietti, A.B., Catalan-Soares, B.C., Murphy, E.L. 2005, Global epidemiology of HTLV-I infection and associated diseases. *Oncogene*. **24**: 6058–6068.

Ravandi, F., Kantarjian, H., Jones, D., Dearden, C., Keating, M., O'Brien, S. 2005, Mature T-cell leukemias. *Cancer*. **104**: 1808–1818.

Semmes, O. J. 2006, Adult T cell leukemia: a tale of two T cells. *J Clin Invest*. **116**: 858–860.

Shuker, S.B., Mariani, V.L., Herger, B.E., Dennison, K.J. 2003, Understanding HTLV-I protease. *Chem Biol*. **10**: 373–380.

Sperka, T., Miklóssy, G., Tie, Y., Bagossi, P., Zahuczy, G., Boross, P., Matúz, K., Harrison, R.W., Weber, I.T., Töszér, J. 2007, Bovine leukemia virus protease: comparison with human T-lymphotropic virus and human immunodeficiency virus proteases. *J Gen Virol*. **88**: 2052–2063.

Takatsuki, K. 2005, Discovery of adult T-cell leukemia. *Retrovirology*. **2**: 16.

Teruya, K., Kawakami, T., Akaji, K., Aimoto, S. 2002, Total synthesis of [L40I, C90A, C109A]-human T-cell leukemia virus type 1 protease. *Tetrahedron Lett*. **43**: 1487–1490.

Tözsér, J., Weber, I.T. 2007, The protease of human T-cell leukemia virus type-1 is a potential therapeutic target. *Curr Pharm Design*. **13**: 1285–1294.

Tözsér, J., Zahuczky, G., Bagossi, P., Louis, J.M., Copeland, T.D., Oroszlan, S., Harrison, R.W., Weber, I.T. 2000, Comparison of the substrate specificity of the human T-cell leukemia virus and human immunodeficiency virus proteinases. *Eur J Biochem*. **267**: 6287–6295.

Ureta-Vidal, A., Angelin-Duclos, C., Tortevoye, P., Murphy, E., Lepère, J.F., Buigues, R.P., Jolly, N., Joubert, M., Carles, G., Pouliquen, J.F., de Thé, G., Moreau, J.P., Gessain, A. 1999, Mother-to-child transmission of human T-cell-leukemia/lymphoma virus type I: implication of high antiviral antibody titer and high proviral load in carrier mothers. *Int J Cancer*. **82**. 832–836.

Vega, S., Kang, L.W., Velazquez-Campoy, A., Kiso, Y., Amzel, L.M., Freire, E. 2004, A structural and thermodynamic escape mechanism from a drug resistant mutation of the HIV-1 protease. *Proteins*. **55:** 594–602.

Wang, Y.-X., Freedberg, D.I., Yamazaki, T., Wingfield, P.T., Stahl, S.J., Kaufman, J.D., Kiso, Y., Torchia, D.A. 1996a, Solution NMR evidence that the HIV-1 protease catalytic aspartyl groups have different ionization states in the complex formed with the asymmetric drug KNI-272. *Biochemistry*. **35**: 9945–9950.

Wang, Y.-X., Freedberg, D.I., Wingfield, P.T., Stahl, S.J., Kaufman, J.D., Kiso, Y., Bhat, T.N., Erickson, J.W., Torchia, D.A. 1996b, Bound water molecules at the interface between the HIV-1 protease and a potent inhibitor, KNI-272, determined by NMR. *J Am Chem Soc*. **118:** 12287–12290.

Yamada, Y., Tomonaga, M., Fukuda, H., Hanada, S., Utsunomiya, A., Tara, M., Sano, M., Ikeda, S., Takatsuki, K., Kozuru, M., Araki, K., Kawano, F., Niimi, M., Tobinai, K., Hotta, T., Shimoyama, M. 2001, A new G-CSF-supported combination chemotherapy, LSG15, for adult T-cell leukaemia-lymphoma: Japan Clinical Oncology Group Study 9303. *Br J Haematol*. **113**: 375–382.

Yamaguchi, K. 1994, Human T-lymphotropic virus type I in Japan. *Lancet*. **343**: 213–216.

Zhang, M., Nguyen, J.-T., Kumada, H.-O., Kimura, T., Cheng, M., Hayashi, Y., Kiso, Y. 2008a, Locking the two ends of tetrapeptidic HTLV-I protease inhibitors inside the enzyme. *Bioorg Med Chem*. **16**: 6880–6890.

Zhang, M., Nguyen, J.-T., Kumada, H.-O., Kimura, T., Cheng, M., Hayashi, Y., Kiso, Y. 2008b, Synthesis and activity of tetrapeptidic HTLV-I protease inhibitors possessing different P_3-cap moieties. *Bioorg Med Chem*. **16**: 5795–5802.

Chapter 6
Picornaviruses

David Neubauer, Jutta Steinberger, and Tim Skern

Abstract The picornavirus family contains several major human and animal pathogens. Vaccines against some of these pathogens are available. However, the availability of potent antiviral compounds would be an appreciable advantage in fighting these pathogens. Inside their non-enveloped capsid, picornaviruses possess a positive sense RNA genome with a single open reading frame. Upon release into the cytoplasm, the genome is translated into a single polyprotein that is processed by virally encoded proteinases. These proteinases represent excellent targets for the development of anti-virals for two reasons. First, efficient polyprotein processing is essential for successful viral replication. Second, the picornaviral proteinases show notable differences to cellular proteinases. To aid in the development of anti-virals, detailed knowledge of the mechanisms, substrate specificities and structures of these proteinases is needed. This chapter reviews recent progress, discusses selected substances with antiviral activity against picornavirus proteinases and outlines several new avenues for the design of novel anti-virals.

Keywords Poliovirus • human rhinovirus • aphthovirus • proteolytic processing • translational control

6.1 Introduction

The family of the positive-strand picornaviruses includes a number of important human and animal pathogens such as poliovirus (PV), hepatitis A virus (HAV), coxsackievirus (CV), human rhinovirus (HRV) and foot-and-mouth disease virus (FMDV) (Racaniello, 2007). Of these pathogens, only PV, HAV and FMDV can be controlled by vaccination. There is thus a clear need for anti-viral agents to combat HRV and CV. In addition, although the vaccines against FMDV and PV have proven effective in most cases, they are not perfect. For instance, the FMDV

D. Neubauer, J. Steinberger, and T. Skern (✉)
Max F. Perutz Laboratories, Medical University of Vienna, Dr. Bohr-Gasse 9/3, A-1030 Vienna, Austria
e-mail: timothy.skern@meduniwien.ac.at

U. Lendeckel and N.M. Hooper (eds.), *Viral Proteases and Antiviral Protease Inhibitor Therapy*, Proteases in Biology and Disease 8,

vaccine can be only used under certain circumstances proscribed by regulatory authorities such as the European Union. Furthermore, the present PV vaccines may not be sufficient to finally achieve and maintain the eradication of PV and may need support from specific anti-viral agents (Aylward et al., 2005). Thus, in addition to the need for anti-virals against HRV and CV, there is also a need for anti-viral compounds against PV and FMDV.

The last 25 years have seen an enormous increase in our knowledge and understanding of the molecular biology and pathogenicity of many family members (see Semler and Wimmer, 2002) and a large number of substances have been shown to possess activity against picornaviruses and/or the proteins that they encode (reviewed in De Palma et al., 2008). Nevertheless, at present no anti-viral substances have been approved for clinical use against picornaviral infections.

This chapter begins by explaining the situations in which an anti-viral against a particular picornavirus would be advantageous, identifies the possible proteolytic activities against which anti-viral substances can be directed as well as reviewing past progress and future directions.

6.1.1 Poliovirus

PV is the subject of a WHO eradication program, initiated in 1988 and originally targeted for completion in 2000 (Robertson et al., 1990). However, the number of cases of wild-type PV reported worldwide in 2007 was 1,310; in the first 5 months of 2008, reports of 354 cases had been received (Anon, 2008). Several reasons for the lack of success in completely eradicating PV are apparent. There is the often low level of immunity induced by the trivalent oral vaccine, a failure to immunise all children, sometimes for religious reasons, and the emergence of recombinant, virulent strains of PV derived from the vaccine strains themselves (Andrianarivelo et al., 1999; Katz, 2006). Furthermore, wild-type strains of PV are still being detected in parts of Africa, even though cases of disease have not been reported (E. Wimmer, personal communication). Finally, certain immune-suppressed patients are capable of shedding PV for many years without showing symptoms (MacLennan et al., 2004; Yoneyama et al., 2001).

Several of the above problems indicate that both wild-type and vaccine-derived PV strains will still be circulating when the WHO eventually recommends the cessation of vaccination against PV. The non-vaccinated population, which would grow every year, would therefore be at risk for infection from any circulating PV. To combat this eventuality, the WHO has made a series of recommendations (Aylward et al., 2005). One recommendation for treating post-vaccination outbreaks of poliomyelitis is the use of an anti-PV agent (Aylward et al., 2005). The suitability and feasibility of anti-viral agents directed against PV was discussed at a meeting convened by the National Academy of Sciences (NAS) of the USA in November 2005. Subsequently, the NAS recommended that such anti-viral agents should be developed to treat post-vaccination outbreaks and to treat immune-comprised patients suffering from persistent infections (N.R.C. Committee on Development of a Polio Antiviral and Its Potential Role in Global Poliomyelitis Eradication, 2006). Further arguments for the development of anti-viral agents against poliovirus can be found in a recent review (Collett et al., 2008).

6.1.2 *Coxsackievirus*

Coxsackieviruses, like PV, also belong to the enterovirus genus of the picornavirus family. Coxsackie B viruses have long been recognised as one of the most significant causes of dilated cardiomyopathy, a major contributor to fatal heart failure in the developed world (Baboonian et al., 1997). One pathway to cell damage during infection is the cleavage of the structural protein dystrophin in myocytes by a virally encoded proteinase (Badorff et al., 1999). This loss of dystrophin also aids the production and release of virus particles (Badorff and Knowlton, 2004). The availability of anti-viral agents to prevent this damage during cardiac surgery and convalescence from heart disease would be of great benefit.

6.1.3 *Human Rhinoviruses*

Human rhinoviruses are the major causative agents of the common cold (Arruda et al., 1997). The replication of these viruses is usually confined to the upper respiratory tract and the illness is usually mild and not dangerous. However, recent investigations have revealed connections to diseases of the lower respiratory tract such as pneumonia and influenza-like disease (Turner, 2007). Furthermore, rhinovirus infections may aggravate the condition of asthma sufferers (Johnston et al., 1993). Two genetic groups of HRVs (HRV-A and HRV-B) have been recognised for some time and found associated with disease (Savolainen et al., 2002). Recently, however, a new genotype of HRV, HRV-C, has been detected in patients suffering from influenza-like disease in which influenza virus could not be detected (McErlean et al., 2007). Viruses from this HRV-C genetic group have now been shown to be causes of hospitalisation in children across the globe (Lau et al., 2007; Renwick et al., 2007). Investigations into the properties of this newly discovered HRV genotype have not yet begun, as it has not been possible to propagate the viruses in cell culture. However, comparison of the available full length genomic sequences shows clearly that these viruses represent a new genetic group. Given the sheer numbers of HRV serotypes, control by vaccination is not an option. However, the availability of inhibitors to treat asthma patients and children hospitalised with respiratory disease brought about by HRV infections would be of significant value in the clinic. This topic has been recently reviewed by Patick (Patick, 2006).

6.1.4 Foot-and-Mouth Disease Virus

Foot-and-mouth disease virus (FMDV), a veterinary pathogen of major economic importance, is still endemic in Africa, Asia and parts of South America (Thomson et al., 2003). The disease is controlled by vaccination and by bans on the movement of infected animals as well as the export of products from infected animals (Grubman and Baxt, 2004). For endemic areas, an anti-viral would not be useful for the simple reasons of cost and the possibility of resistance. There are however two scenarios in which an anti-viral agent would be useful for the treatment or prevention of FMDV infections. These are in the treatment of outbreaks in areas which have been free of FMDV (e.g. in England in 2007) and in the treatment of persistent infections. Persistently infected animals are a problem in Africa, affecting buffaloes and impala (Thomson et al., 2003). The virus excreted by such animals can be transmitted to local farm stock and cause difficulty in the sale and movement of these animals. The availability of anti-viral agents against FMDV would make it possible to test whether persistently infected animals can be cured of the infection.

6.1.5 Hepatitis A virus

Since the 1990s, control of the single serotype HAV has been possible using an inactivated vaccine that induces a strong and protective antibody response. Nevertheless, in 2003, over 7,000 reported cases of HAV were reported in the USA alone, the majority being adults (Brundage and Fitzpatrick, 2006). There is thus still a need for the development of an anti-viral to shorten the period of illness, mitigate its symptoms and reduce the time of convalescence. An anti-viral would be of appreciable importance to those patients who suffer from relapses as well as the 1% of patients suffering liver failure and requiring a liver transplant (Brundage and Fitzpatrick, 2006).

6.1.6 Proteolytic Targets for Anti-viral Compounds in Picornaviruses

The above examples illustrate the need for further measures to control these members of the picornavirus family. In order to discuss the potential targets for anti-viral agents encoded by the picornaviruses, it is necessary to consider how the information present in the picornavirus genome is expressed. Picornaviruses possess a genome of positive polarity which is surrounded by a non-enveloped capsid comprising 60 copies of each of the four capsid proteins, VP1–VP4. The information in the genome is first expressed as a large polyprotein which is cleaved

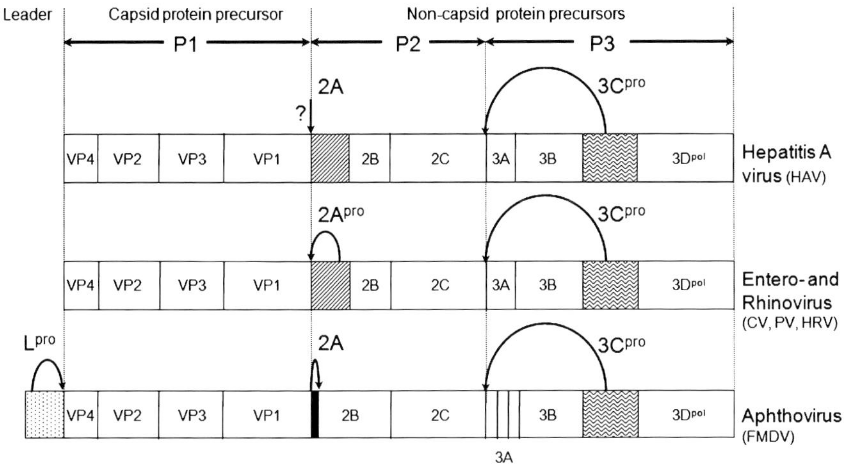

Fig. 6.1 Proteolytic processing in picornaviruses. Variations in primary cleavage events in four picornavirus genera. The polyproteins of the indicated viruses are shown schematically as open boxes. Primary cleavages are indicated. The different shadings of the 2A protein reflect differences in mechanism and size in this protein between the different genera. The proteolytic agent separating VP1 from 2A in HAV is believed to be an as yet unidentified host cell protease (Morace et al., 2008) (Adapted from Skern et al., 2002)

into the mature viral proteins by virally encoded proteinases (Racaniello, 2007; Fig. 6.1). Hepatoviruses (HAV) encode just one proteinase whereas enteroviruses (PV and CV), rhinoviruses and aphthoviruses (FMDV) all encode two proteinases for processing. The 3C protease ($3C^{pro}$), a chymotrypsin-like cysteine proteinase responsible for the majority of cleavage events, is found in all picornaviruses. The second proteinase encoded by entero- and rhinoviruses, the 2A protease ($2A^{pro}$), is also a cysteine proteinase with a chymotrypsin-like fold. In contrast, the second proteinase encoded by FMDV, the Leader protease (L^{pro}), is a papain-like cysteine proteinase (Skern et al., 2002). All three of these proteinases differ in many ways from the proteinases of the host and consequently represent drug targets. The 2A protein of FMDV has, in contrast, no proteolytic activity. Instead, the RNA sequence encoding the last three amino acids of the 2A interrupts the synthesis of the polyprotein chain. This particular sequence on the RNA leads the ribosome to pause and release the structural precursor. However, a certain percentage of the ribosomes remain attached to the RNA and continue translation without changing the reading frame. This allows the non-structural part of the polyprotein to be translated, but at reduced levels compared to the structural precursor. This mechanism is referred to as a ribosome skip (Atkins et al., 2007). A third target for anti-viral agents in all picornaviruses is, of course, the viral RNA polymerase ($3D^{pol}$), the enzyme replicating the viral RNA for which there is no cellular counterpart.

Over the years, most attention has focussed on the $3C^{pro}$ and $3D^{pol}$ enzymes because they are common to all picornaviruses and are involved in many steps of processing and replication. Indeed, a large body of information on the structures

Table 6.1 Overview of known structures of picornaviral proteases

Protease	Virus	Method	Resolution	PDB-entry	Reference
$3C^{pro}$	HRV2	X-Ray	1.85Å	1CQQ	Matthews et al. (1999)
	HRV14	X-Ray	2.30Å	Not available	Matthews et al. (1994)
	HRV14	NMR		2IN2	Bjorndahl et al. (2007)
	PV1	X-Ray	2.10Å	1L1N	Mosimann et al. (1997)
	HAV	X-Ray	2.00Å	1HAV	Bergmann et al. (1997)
	FMDV	X-Ray	1.90Å	2BHG	Birtley et al. (2005)
	CVB3	X-Ray	2.40Å	2VB0	Anand et al., unpublished
$3CD^{pro}$	PV	X-Ray	3.00Å	2IJD	Marcotte et al. (2007)
$2A^{pro}$	HRV2	X-Ray	1.95Å	2HRV	Petersen et al. (1999)
	CVB4	NMR		1Z8R	Baxter et al. (2006)
L^{pro}	FMDV	X-Ray	3.00Å	1QOL	Guarné et al. (1998)
	FMDV	NMR		2JQF	Cencic et al. (2007)
sL^{pro}	FMDV	NMR		2JQG	Cencic et al. (2007)

and biochemistry of these enzymes is available (for $3C^{pro}$ see Table 6.1; for $3D^{pol}$ see Ferrer-Orta et al., 2007; Hansen et al., 1997; Lyle et al., 2002; Marcotte et al., 2007). Furthermore, inhibitors to both enzymes have been developed (Dragovich et al., 1998a; Harki et al., 2006; Huitema et al., 2008) and at least one $3C^{pro}$ inhibitor (AG-7088, also known as rupintrivir) has been tested in the clinic (Hayden et al., 2003; Witherell, 2000). The structures of picornaviral proteinases that have been determined are summarised in Table 6.1.

Experience with anti-viral agents against human immunodeficiency virus (HIV) has shown that it is important to have more than one target for anti-viral agents to combat the development of resistance in viruses with RNA genomes. Nevertheless, in contrast to the situation with $3C^{pro}$ and $3D^{pol}$, interest in the $2A^{pro}$ of entero- and rhinoviruses and the Lb^{pro} of FMDV as drug targets has been limited. This lack of interest is based on the fact that both $2A^{pro}$ and Lb^{pro} perform just one single intramolecular cleavage on their respective polyprotein. Given the rapid kinetics of these reactions (Glaser et al., 2001; Glaser et al., 2003), the general consensus of opinion has been that it will be difficult to inhibit these reactions in the infected cell. However, recent work by Crowder and Kirkegaard (Crowder and Kirkegaard, 2005) has suggested that the inhibition of $2A^{pro}$ cleavage by anti-viral agents would actually be a very effective strategy to block replication of those picornaviruses encoding such an enzyme. Crowder and Kirkegaard showed that mutations in $2A^{pro}$ had a trans dominant effect on the replication of the wild-type virus. Thus, co-transfection of a wild-type PV RNA with a PV RNA containing a debilitating mutation in $2A^{pro}$ led to a reduction in the replication of the wild-type virus. The simplest explanation of this result is that the mutant virus fails to free the $2A^{pro}$ from the capsid protein precursor. This capsid precursor with the $2A^{pro}$ extension can also be incorporated into the assembling wild-type capsid; however, the $2A^{pro}$ extension prevents completion of the capsid and thus has a detrimental effect on the assembly and virus production of the wild-type. This phenomenon has two implications for the targeting of the $2A^{pro}$. First, it is not necessary for the $2A^{pro}$ to be inhibited completely. A few incorrect capsid precursors will be able to interfere with a large number of correctly processed ones. Second, it will be theoretically more difficult for the virus to develop resistance to the inhibitor because the presence of sensitive viruses will upset the replication of any resistant mutants which may arise (Crowder and Kirkegaard, 2005; Semler, 2005).

A similar situation can be imagined in FMDV with mutants that prevent processing by the Lb^{pro}. As mentioned above, Lb^{pro} also just carries out one intramolecular cleavage on the viral protein; unlike $2A^{pro}$, this cleavage is between its own C-terminus and the N-terminus of the capsid protein precursor (Fig. 6.1). This will eventually lead to the generation of an N-terminal extension of VP4. Incorporation of such molecules into the capsid can be expected to have a detrimental effect on viral infectivity similar to that found with the $2A^{pro}$ extension of VP1 in PV.

In order to make use of the observation of Crowder and Kirkegaard (Crowder and Kirkegaard, 2005), a thorough knowledge of the molecular mechanisms of the $2A^{pro}$ and Lb^{pro} is required. Over the years, a substantial amount of biochemical and structural information on these enzymes has been generated (Table 6.1; reviewed in Skern et al., 2002; Baxter et al., 2006; Cencic et al., 2007). However, much knowledge of fundamental importance, such as how $2A^{pro}$ specificity is defined and whether there is a difference between the intra- and intermolecular cleavage reactions, remains outstanding.

6.2 Picornaviral 3C Proteinases

The first investigations of the protein processing cascade in picornaviruses used inhibitor profiling to identify $3C^{pro}$ as a member of the cysteine protease family (Gorbalenya and Svitkin, 1983; Pelham, 1978). Subsequently, comparison of protein sequences and protein modelling revealed that the $3C^{pro}$, although cysteine proteases, possess a similar fold to that of chymotrypsin (Argos, 1984; Bazan and Fletterick, 1988; Gorbalenya et al., 1989). The catalytic triad of the $3C^{pro}$ was, however, predicted to comprise the residues cysteine, histidine and glutamate (Gorbalenya et al., 1989). The first structures of the $3C^{pro}$ of HAV (Allaire et al., 1994) and HRV14 (Matthews et al., 1994) confirmed these predictions of a chymotrypsin fold. Furthermore, the catalytic triad of the HRV14 $3C^{pro}$ did indeed have glutamate as the third residue of the active site. The situation in the HAV enzyme was however less clear. Initially, James and co-workers favoured the hypothesis that residue tyrosine 143 oriented the active site histidine residue (Allaire et al., 1994; Bergmann et al., 1997). However, in later work using an alternative crystal form, the active site had rearranged to allow aspartate 84 to orient the histidine, as seen in the canonical chymotrypsin-like enzymes (Yin et al., 2005). Subsequent structures of $3C^{pro}$ from other genera revealed that the third member of the catalytic triad was glutamate in PV $3C^{pro}$ and aspartate in the FMDV enzyme (Birtley et al., 2005; Mosimann et al., 1997).

The $3C^{pro}$ are the workhorses of the picornaviral processing reaction. Starting at the 2C/3A junction, they cleave to generate all the mature proteins shown in Fig. 6.1, except at the site between VP4 and VP2 and at the sites mentioned above for which the L and 2A proteinases are responsible. The VP4/VP2 cleavage takes place during maturation of the viral capsid by an as yet unidentified proteolytic activity.

The sequences at which the $3C^{pro}$ cleave vary between the genera. The most specific enzymes are the PV and HRV $3C^{pro}$, requiring a small residue at P4, glutamine at P1 and glycine at P1′. The HRV $3C^{pro}$ in particular also shows a strong preference for proline and other hydrophobic residues at the P2′ position (Duechler et al., 1987; Skern et al., 2002). In contrast, the HAV $3C^{pro}$ has a preference for a bulky, hydrophobic residue at P4, a small aliphatic residue at P2 and glutamine or glutamate at P1. The HAV $3C^{pro}$ has little or no specificity at the P′ side (Bergmann et al., 1997). Like the HAV enzyme, the FMDV $3C^{pro}$ can also accept glutamine or glutamate at P1 and prefers a bulky residue at P4. The P2 residue is frequently lysine or threonine. At the P1′ position, the enzyme can accept glycine, serine and a variety of large hydrophobic residues (Birtley et al., 2005). The structural basis for these cleavage requirements are briefly explained for the individual enzymes in the sections below.

In addition to the processing on the viral polyprotein, picornaviral $3C^{pro}$ have been shown to cleave host cell proteins during viral replication; many of these cleavages modulate transcription and translation in the infected cell (reviewed in Lloyd, 2006). However, at present, no common cellular target for all picornaviral $3C^{pro}$ has been identified. For this reason, the cellular targets of $3C^{pro}$ are mentioned in the specific sections below devoted to the $3C^{pro}$ of the different genera.

In addition to all these proteolytic activities, all the picornaviral $3C^{pro}$ also possess an RNA binding site, located on the opposite face of the molecule to that responsible for proteolysis (Allaire et al., 1994; Matthews et al., 1994; Skern et al., 2002). Specifically, the RNA binding site comprises parts of the N- and C-terminal helices as well as the part of the polypeptide chain that links the N- and C-domains. In entero- and rhinoviruses, this RNA binding site of $3C^{pro}$ or its precursor 3CD has been shown to bind to the clover-leaf structure at the 5′ end of the viral RNA to set up the replication complex on the genomic RNA (Andino et al., 1990; Zell et al., 2002). As this RNA binding sequence is extremely well conserved throughout the picornaviruses (Skern et al., 2002; Yin et al., 2005), a substance that interferes with the RNA binding site of one picornaviral $3C^{pro}$ may interfere with several other viruses across the different genera.

Comprehensive lists and descriptions of all $3C^{pro}$ inhibitors developed in the recent years have been compiled by Lall (Lall et al., 2004) and De Palma (De Palma et al., 2008).

6.2.1 Poliovirus $3C^{pro}$

The PV $3C^{pro}$ is the most specific of its class, cleaving solely at glutamine-glycine amino acid pairs. The structure of this $3C^{pro}$, determined by Mosimann et al. (Mosimann et al., 1997), revealed that glycine at the P1′ was required to turn the polypeptide chain of the substrate away from the β-strand bI1 that effectively prevents the acceptance of any amino acid with a side-chain at the P1′ position. The specificity for the glutamine at P1 appears to be due to the presence of the uncharged residue histidine 161 that lies at the bottom of the P1 pocket. Importantly, histidine 161 is maintained in an uncharged state by hydrogen bonding of the hydroxyl group of tyrosine 138 with the nitrogen atom in the imidazole ring (Mosimann et al., 1997).

The specificity of PV $3C^{pro}$ whilst still bound to the $3D^{pol}$ protein is different from that of the mature $3C^{pro}$. Ympa-Wong et al. (Ypma-Wong et al., 1988) have shown that the 3CD protein is the form of $3C^{pro}$ responsible for cleaving the cleavages sites between capsid proteins VP2 and VP3 and between VP3 and VP1. However, the molecular and structural basis for this observation is not known. Even the recent determination of the structure of the 3CD precursor (Marcotte et al., 2007) failed to shed light on this issue.

Poliovirus $3C^{pro}$ has been reported to cleave a large variety of cellular proteins involved in transcription (Skern et al., 2002), translation (Lloyd, 2006; Perera et al., 2007; de Breyne, 2008), the ultrastructure of the cell (Joachims et al., 1995) and inflammation (Neznanov et al., 2005).

Sarkany and Polgar (Sarkany and Polgar, 2003) have examined in detail the catalytic mechanism of PV $3C^{pro}$ and note several unusual aspects whilst Gouvea and colleagues noted an unexpected behaviour of the enzyme towards chaotropic salts (Gouvea et al., 2006).

Given the need for a PV anti-viral outlined above, it is suprising that there are very few reports documenting specific inhibitors of PV $3C^{pro}$, in contrast to the situation with HRV (De Palma et al., 2008) and HAV (Lall et al., 2004).

6.2.2 *Rhinovirus $3C^{pro}$*

The groups of Matthews and Patick have made an extensive study of HRV $3C^{pro}$ and have developed many compounds active against these enzymes (Dragovich et al., 1998b; Hayden et al., 2003; Matthews et al., 1999; Matthews et al., 1994). One of the most effective compounds was AG7088 (later known as rupintrivir) (Matthews et al., 1999), a potent inhibitor of all HRV serotypes tested as well as some other related human enteroviruses (Binford et al., 2005). This compound showed promise in a phase II trial in the clinic (Hayden et al., 2003); however, in a further trial, the compound failed to diminish both viral load and disease severity and was thus not developed further (Patick, 2006; Patick et al., 2005).

The development of rupintrivir (shown bound to HRV2 $3C^{pro}$ in Fig. 6.2a) clearly illustrates that it is possible to develop a single inhibitor that is capable of inhibiting the replication of most rhinovirus serotypes (Binford et al., 2005). Such a property is not immediately obvious as the identity between serotypes can be as low as 50% (Duechler et al., 1987; Skern et al., 1985). However, analysis shows that many of the $3C^{pro}$ residues involved in binding rupintrivir are conserved in HRVs and human

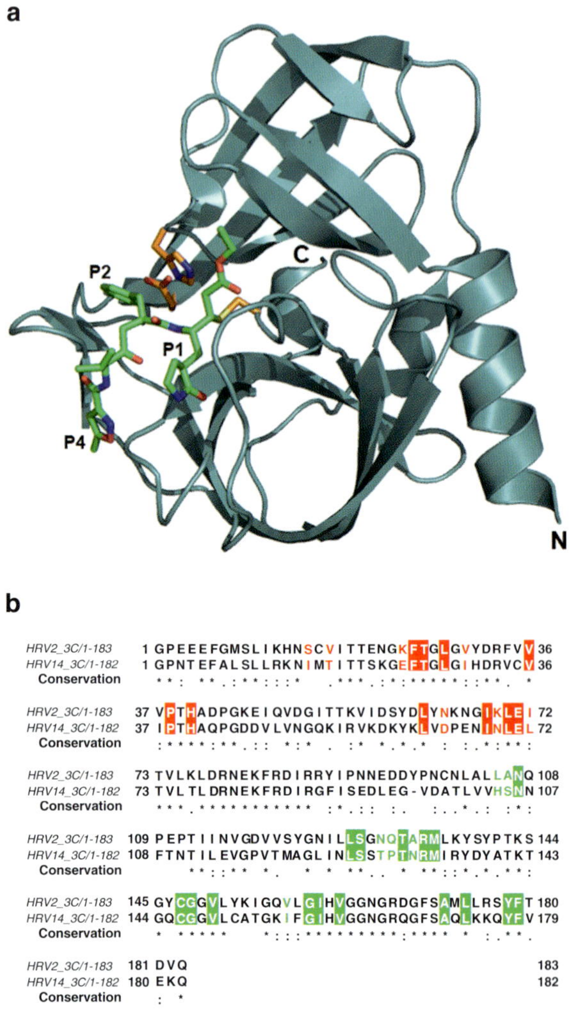

Fig. 6.2 The structure of HRV2 3C[pro] (PDB 1CQQ) bound to the inhibitor rupintrivir and amino acid sequence conservation at the domain-interface. (**a**) Crystal structure of the HRV2 3C[pro] covalently bound to the inhibitor rupintrivir. The protein is shown as a cartoon in blue with its catalytic triad shown as orange sticks. The inhibitor is depicted as a stick model, with carbon atoms in green, oxygen atoms in red, sulphur atoms in yellow, and nitrogen atoms in blue. The groups of the inhibitor mimicking the P4, P2 and P1 sites are labeled (generated with PyMOL DeLano, 2002). (**b**) Amino acid sequence alignment of HRV2 and HRV14 3C[pro]. Residues lining the interface between the N- and C-terminal domain are colored red and green respectively. Conserved residues are shown in reverse video (generated with JalView Clamp et al., 2004)

enteroviruses, including PV (Binford et al., 2005). Interestingly, rupintrivir has been shown to be very effective against PV in cell culture (De Palma, communication at the XIV Meeting of the European Study Group on the Molecular Biology of Picornaviruses, Finland, Nov. 2006). In addition, Fig. 6.2a shows that rupintrivir makes contacts to both the N-terminal and C-terminal domains. Figure 6.2b shows that most of the residues at this domain interface are indeed conserved. As mentioned below, this is not the case for the domain interfaces of HRV 2A[pro].

What is the basis for the potency of rupintrivir? The electron acceptor of rupintrivir is a Michael group which forms a stable covalent adduct with the SH group of the active site nucleophile cysteine 147 of the 3C[pro]. In Fig. 6.2a, the cysteine 147 side-chain is covalently attached to the inhibitor. The specificity of rupintrivir derives from the moieties occupying the P1, P2 and P4 positions. At P1, a lactam ring (actually a cyclised analogue of glutamine) is present that makes favourable contacts with the residues of the P1 pocket (Fig. 6.2a) and may also stablise the free inhibitor in a conformation close to that found in the bound inhibitor (Matthews et al., 1999). The P2 residue of the inhibitor is a fluoro-phenyl moiety that mimics the large, bulky residue that is preferred by HRV $3C^{pro}$. Finally, the P4 moiety of the inhibitor is an iso-oxazole derivative that is well accepted by the small P4 pocket of these enzymes (Matthews et al., 1999). The position of each of the three moieties can be clearly seen in Fig. 6.2a.

Like PV $3C^{pro}$, HRV14 $3C^{pro}$ has been shown to cleave the p65-RelA subunit of NF-kappaB during replication (Neznanov et al., 2005).

6.2.3 Coxsackievirus $3C^{pro}$

In contrast to the other $3C^{pro}$, the first structural work on CV $3C^{pro}$ was to investigate its interaction with a synthetic RNA representing one of the stem-loops from the clover-leaf at the 5′ end of the viral RNA (Ohlenschlager et al., 2004; Zell et al., 2002). Interestingly, these observations show that the $3C^{pro}$ binds to a specific structure rather than a particular sequence of nucleotides. This suggests that inhibitors of CV $3C^{pro}$ binding may also inhibit the RNA binding of $3C^{pro}$ across the genera. Very recently, the structure of the CV $3C^{pro}$ itself was determined by Anand and co-workers using X-ray crystallography and the co-ordinates deposited in the database (see Table 6.1). This structure should aid in developing specific inhibitors for the CV $3C^{pro}$ and in understanding the interactions between the protein and the 5′ RNA clover-leaf.

Lee et al. have used rupintrivir as a lead compound to develop compounds effective against CVB2 $3C^{pro}$ (Lee et al., 2007).

6.2.4 Hepatitis A virus $3C^{pro}$

HAV $3C^{pro}$ differs strongly from the rhino- and enterovirus enzymes in that it prefers large residues at P4, smaller ones at P2, can accept glutamine or glutamate at P1 and has no specificity at the P' side. It can even accept such bulky residues as methionine or arginine at the P1′ position (Skern et al., 2002). Specificity at P4 in HAV $3C^{pro}$ is achieved through a larger pocket than in PV $3C^{pro}$; at P2, a small pocket is present for this residue that is absent in PV $3C^{pro}$. The P1 pocket residue of HAV is, as in HRV and PV $3C^{pro}$, an uncharged histidine residue, although the residues orienting this residue are not conserved (Skern et al., 2002). The difference at the P′ side is explained by the greater distance of the β-strand bI1 from the active site in the HAV $3C^{pro}$, removing the need for a glycine residue to turn the polypeptide chain of the substrate from this structure (Bergmann et al., 1997).

Several reports have described specific inhibitors for HAV $3C^{pro}$ (reviewed in Huitema et al., 2008; Lall et al., 2004; Yin et al., 2005). The enzyme has also been shown to cleave the host cell proteins poly(A) binding protein (PABP) and poly(C) binding protein II (Zhang et al., 2007a, b). Another unusual characteristic of this enzyme is that it has been reported to bind structures from the 5′ region of the viral RNA much more efficiently as a dimer than as a monomer (Peters et al., 2005).

6.2.5 *Foot-and-Mouth Disease $3C^{pro}$*

The structure of the $3C^{pro}$ of FMDV is similar to those of other picornaviral $3C^{pro}$, the major difference being the flexibility of a β-ribbon in the C-terminus of the molecule. This ribbon is capable of folding over the substrate binding site of the enzyme and providing residues involved in determining specificity. For example, the β-ribbon residue cysteine 142 appears to be involved in binding to the generally hydrophobic side chains of the P4 and P2 residues (Curry et al., 2007; Sweeney et al., 2007). The unusual ability of this enzyme to accept glutamate or glutamine equally at P1 remains unclear as the arrangement of the FMDV $3C^{pro}$ pocket accepting the P1 residue does not differ greatly from those of $3C^{pro}$ that discriminate against glutamate (e.g. HRV2 $3C^{pro}$ and PV $3C^{pro}$) (Birtley et al., 2005).

The FMDV $3C^{pro}$ has been shown to cleave a number of cellular proteins, including histone 3C (Falk et al., 1990; Tesar and Marquardt, 1990), the translation factors eIF4A (Li et al., 2001a) and eIF4G (Belsham et al., 2000) and gamma-tubulin (Armer et al., 2008).

6.3 Enteroviral and Rhinoviral 2A Proteinases

The first report that the 2A protein of PV contained a proteolytic activity cleaving between the C-terminus of VP1 and its own N-terminus was published by Toyoda et al. (Toyoda et al., 1986) in 1986. Subsequently, the same activity was demonstrated for the 2A proteins of HRV and CV (Liebig et al., 1993; Sommergruber et al., 1989). Inhibitor profiling and protein modelling showed that the $2A^{pro}$, like the $3C^{pro}$, are cysteine proteinases with a chymotrypsin-like fold. It seems likely that the $2A^{pro}$ arose by a duplication of the $3C^{pro}$ on the evolutionary precursor of the entero- and rhinoviruses. Despite this close relationship, there are however several clear differences between the $2A^{pro}$ and $3C^{pro}$ that preclude development of a single inhibitor for both proteinases. These differences concern the overall structures, the mechanisms and the specificity determinants. Thus, although the $2A^{pro}$ are closely related to the smaller serine proteinases such as streptomyces G protease B (SGPB) and α-lytic proteinase (Bazan and Fletterick, 1988; Petersen et al., 1999), structural analysis of the HRV and CV $2A^{pro}$ shows that their N-terminal domain lacks four β-strands present in the N-terminal domain of SGPB and $3C^{pro}$. In contrast, the

$2A^{pro}$ contain a zinc ion in the C-terminal domain, using a structural motif that is unique amongst the chymotrypsin-like proteinases (Petersen et al., 1999). In terms of mechanism, all $2A^{pro}$ use an aspartate residue as the third member of their active site triad; as mentioned above, glutamate or aspartate maintain this function in the $3C^{pro}$. Turning to specificity, the main specificity determinants for the $2A^{pro}$ are at P4, P2, P1′ and P2′, with several residues being accepted at the P1 position (Skern et al., 2002). This is in clear contrast to the well-defined P1 specificities observed in the $3C^{pro}$.

Rhino- and enteroviral $2A^{pro}$ have also been shown to cleave host cell proteins. In contrast to $3C^{pro}$, however, all $2A^{pro}$ have one cellular target in common, namely the cellular translation molecule eukaryotic initiation factor (eIF) 4G (Morley et al., 1997). This protein is present as two homologues, eIF4GI and eIF4GII (Gradi et al., 1998a). Cleavage of these homologues at a single site leads to the inability of the host cell to synthesise protein from its own capped mRNA. In contrast, viral RNA can still be translated from its internal ribosome entry site (IRES) and is even stimulated under these conditions (Ziegler et al., 1995). Individual $2A^{pro}$ have also been shown to cleave a variety of other cellular proteins. Some of these are discussed in the sections on the individual $2A^{pro}$ below.

The available structures for rhino- and enteroviral $2A^{pro}$ are listed in Table 6.1.

6.3.1 Poliovirus $2A^{pro}$

The PV $2A^{pro}$ was purified to homogeneity from infected cells just 2 years after its identification as a protease (Koenig, 1988). Subsequently, the recombinant protein was purified by affinity chromatography using maltose-binding protein and hexahistidine tags (Ventoso et al., 1998; Yalamanchili et al., 1997). In spite of these successes, it has not been possible to produce material of sufficient quantity and purity to allow structural work to proceed. This structure is perhaps the most important remaining target amongst the picornaviral proteases.

In contrast, much biochemical and molecular biological work has revealed a multitude of proteins that are cleaved by PV $2A^{pro}$. In addition to the eIF4G homologues, these cellular targets of PV $2A^{pro}$ include amongst others: the poly(A) binding protein PABP (Joachims et al., 1999), the TATA-binding protein (Yalamanchili et al., 1997), the catalytic subunit of the DNA-dependent protein kinase (Graham et al., 2004), proteins of the nuclear pore complex (Gustin and Sarnow, 2001; Park et al., 2008) and the protein gemin of the U snRNP assembly (Almstead and Sarnow, 2007).

The above list documents that the PV $2A^{pro}$ plays a major role in tailoring the infected cell to the needs of the virus. In addition, over and above its role in proteolytic processing, the PV $2A^{pro}$ has been shown to be involved in regulating stability, replication and translation of the RNA (Jurgens et al., 2006; Li et al., 2001b). PV $2A^{pro}$ is therefore truly a multifunctional enzyme. Along with the trans-dominant effect of certain $2A^{pro}$ mutations referred to above (Crowder and Kirkegaard, 2005), these properties make PV $2A^{pro}$ an excellent target for anti-viral substances.

6.3.2 *Human Rhinovirus 2A^pro^*

As mentioned above, the HRV genus contains over 100 antigenically distinct serotypes that can be grouped into the genetic clusters A and B. Comparison of the $2A^{pro}$ sequences from members of the different groups reveals an identity of only 40% (Sousa et al., 2006), about 10% less than between the $3C^{pro}$ of the different clusters (Argos, 1984). Nevertheless, for the $3C^{pro}$, these differences did not prevent the development of an inhibitor such as rupintrivir that was capable of inhibiting the $3C^{pro}$ of both groups and of several enteroviruses (De Palma et al., 2008). For HRV $2A^{pro}$, however, several lines of evidence suggest that the differences between the genetic group A and B $2A^{pro}$ will be sufficient to impede the development of a general inhibitor for HRV $2A^{pro}$. Indeed, over the years, we and others have made several observations that indicate differences in specificity and possibly also in mechanism between the $2A^{pro}$ of various rhinovirus and enterovirus $2A^{pro}$. These observations are summarised below.

First, the specificities of the HRV2 and HRV14 $2A^{pro}$, although not fully understood, appear to be different. Table 6.2 shows that the self-processing cleavage sites for the two enzymes clearly have only a few residues in common. Table 6.2 also shows the cleavage sites of HRV2 $2A^{pro}$ on the eIF4G homologues; it is believed that HRV14 $2A^{pro}$ cleaves the eIF4G homologues at these sites but it has not yet been demonstrated to do so. We have investigated this question by replacing the self-processing cleavage site of both HRV2 $2A^{pro}$ and HRV14 $2A^{pro}$ with that of the eIF4GI sequence shown in Table 6.2. The $2A^{pro}$ of HRV2 cleaved the eIF4GI site with the same efficiency as the wild-type sequence. In contrast, the eIF4GI cleavage site was refractory to HRV14 $2A^{pro}$ cleavage (Sousa et al., 2006). Using site-directed mutagenesis, we determined that lack of cleavage by the HRV14 enzyme was due to the presence of arginine at the P1 site in the self-processing reaction. HRV14 $2A^{pro}$ could not accept this residue whereas the HRV2 enzyme could (Sousa et al., 2006). This suggests that the substrate binding pockets for the P1 residue on these two $2A^{pro}$ differ considerably.

To examine this question more carefully, we examined the residues proposed to be involved in binding the P1 in HRV2 $2A^{pro}$, the only $2A^{pro}$ for which a high-resolution structure is available. In HRV2 $2A^{pro}$, the residue at the bottom of the P1 pocket appears to be Cys101 (Petersen et al., 1999); the corresponding residue in HRV14 would be A104. Substitution of A104 with cysteine did not, however, confer the

Table 6.2 Known cleavage sites of HRV2 and HRV14 $2A^{pro}$. The arrow indicates the cleavable bond. The cleavage sites on the eIF4G homologues have only been determined for HRV2 $2A^{pro}$ (Gradi et al., 2003; Skern et al., 2002)

Substrate	P5	P4	P3	P2	P1	↓	P1′	P2′	P3′	P4′
HRV2 (VP1–$2A^{pro}$)	Ile	Ile	Thr	Thr	Ala		Gly	Pro	Ser	Asp
HRV14 (VP1–$2A^{pro}$)	Asp	Ile	Lys	Ser	Tyr		Gly	Leu	Gly	Pro
eIF4G I	Thr	Leu	Ser	Thr	Arg		Gly	Pro	Pro	Arg
eIF4G II	Pro	Leu	Leu	Asn	Val		Gly	Ser	Arg	Arg

ability to recognise arginine at P1 by the HRV14 $2A^{pro}$ (Sousa et al., 2006). These results show that substrate recognition differs between enzymes of the two genetic groups and indicate that we do not at present know exactly which $2A^{pro}$ residues are involved in recognising the P1 residue. Insight into this question will require the determination of one or more structures of a $2A^{pro}$ from the genetic group B as well as a structure of a $2A^{pro}$ bound to an inhibitor or to a substrate analogue.

Another clear difference between HRV2 and HRV14 $2A^{pro}$ is their behaviour toward the inhibitor zVAM.fmk (benzyloxycarbonyl-Val-Ala-Met-fluoromethylketone). This inhibitor was developed after the observation by Deszcz et al. (Deszcz et al., 2004) that benzyloxycarbonyl-Val-Ala-Asp (OMe)-fluoromethylketone (zVAD.fmk) could inhibit replication of HRV2 by inhibiting the activity of the HRV2 $2A^{pro}$. zVAD.fmk was originally designed as a caspase inhibitor and is synthesised in an uncharged form to allow passage of the inhibitor through the cell membrane. Inside the cell, the methyl group on the aspartic acid residue is removed and the compound is able to inactivate caspases. Deszcz et al. (Deszcz et al., 2004) determined that the uncharged form of the inhibitor was responsible for the inhibition of the $2A^{pro}$. This agreed with previous data that had indicated that HRV2 $2A^{pro}$ can accept basic and hydrophobic residues at P1 but not acidic residues (Skern et al., 1991). Deszcz and colleagues (Deszcz et al., 2006) made use of this property to synthesise zVAM.fmk. This substance should inhibit HRV2 $2A^{pro}$ through the hydrophobic methionine residue but cannot be activated to an inhibitor of caspases. In cell culture experiments, zVAM.fmk could inhibit the replication not only of the genetic group A viruses HRV2 and HRV16, but also of the genetic group B virus HRV14 (Deszcz et al., 2006). This suggested a common mechanism of inhibition. However, further analysis showed that the HRV14 $2A^{pro}$ is inhibited in both intra- and intermolecular cleavage by zVAM.fmk whereas the HRV2 $2A^{pro}$ is only inhibited in intermolecular cleavage (Deszcz et al., 2006) (Sousa, C. and Skern, T., 2007, unpublished). Although these experiments indicate differences in HRV $2A^{pro}$, they do suggest that it may be possible to find inhibitors that inactivate a spectrum of $2A^{pro}$, even if the mechanism of inhibition differs. Second, these experiments clearly show that it is possible to design inhibitors that are specific for the $2A^{pro}$ but do not possess anti-caspase activity.

Another approach to finding compounds that inactivate the replication of a broad spectrum of HRVs would be to target genetic group A and B HRV separately. The above results suggest that a derivative of zVAM.fmk in which the methionine at the P1 position is replaced with arginine (zVAR.fmk, benzyloxycarbonyl-Val-Ala-Arg-fluoromethylketone) will inhibit the replication of genetic A group viruses but not genetic B group viruses. If this is true, it may be possible to specifically target genetic group B HRV in the same way.

A third example of differences between the genetic group A and B HRV is the difference in the onset of cleavage of eIF4GI and eIF4GII observed in cell culture (Gradi et al., 1998b; Seipelt et al., 2000; Svitkin et al., 1999). HRV2 $2A^{pro}$ has been shown to cleave the two homologues at about the same time during replication whereas the HRV14 $2A^{pro}$ clearly cleaves eIF4GI before eIF4GII. Neither the biological relevance nor the basis of this difference is understood,

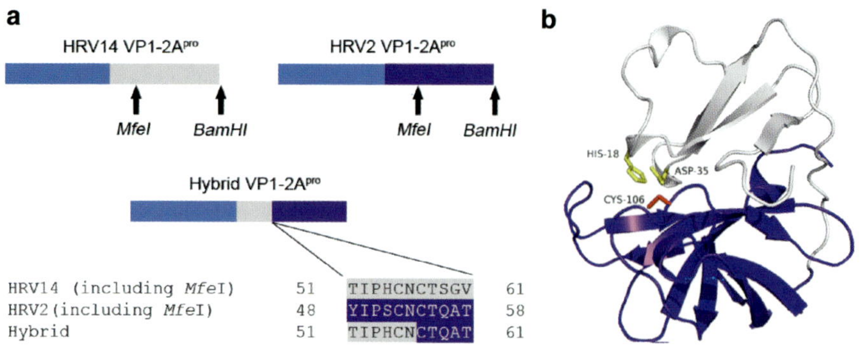

Fig. 6.3 Construction of an HRV14/2 hybrid 2A^{pro}. (**a**) *Mfe*I sites were introduced into the plasmids coding for either HRV2 or HRV14 VP1–2A^{pro} to enable exchange of restriction enzyme fragments encompassing an intact N- or C-terminal domain. An *Mfe*I and *Bam*HI fragment of HRV2 2A^{pro} was used to replace the C-terminal domain of HRV14 2A^{pro} and create a hybrid containing HRV14 VP1, the N-terminal 2A^{pro} domain of HRV14 and the C-terminal 2A^{pro} domain of HRV2. The sequence at the changeover is given. (**b**) The structure of HRV2 2A^{pro} (2HRV) is coloured according to its N-terminal and C-terminal domain. In the hybrid 2A^{pro}, the grey N-terminal domain will be exchanged by the respective HRV14 2A^{pro} domain. The catalytic triad (numbered according to HRV2 2A^{pro}) is coloured in yellow (His18 and Asp35) and red (Cys106) (generated with PyMOL DeLano, 2002)

although the difference could be a reflection of the above-mentioned differences in substrate specificity and inhibitor sensitivity.

The above examples document three phenotypic differences between the HRV2 and HRV14 2A^{pro}. We hypothesised that phenotypic differences might be determined by the property of one of the two domains of the 2A^{pro}. In an attempt to investigate this further, we constructed a hybrid 2A^{pro} containing the N-terminal domain of HRV14 and the C-terminal domain of HRV2. To achieve this, we noticed that an *Mfe*I restriction site could be introduced into the coding region of both enzymes at two of the four residues that form the zinc binding site. These two residues, asparagine 56 and cysteine 57 in HRV14 and asparagine 53 and cysteine 54 in HRV2 2A^{pro}, lie at the C-terminal end of the polypeptide linking the two domains (Fig. 6.3). Using the *Mfe*I and *Bam*HI sites, we introduced the C-terminal domain of HRV2 2A^{pro} (Fig. 6.3a, dark blue) into the HRV14 2A^{pro} (Fig. 6.3a, grey). Figure 6.3b shows the domain arrangement using the same colour scheme, based on the structure of HRV2 2A^{pro}.

We then tested the ability of this hybrid 2A^{pro} to carry out the self-processing reaction at the C-terminus of VP1 and its own N-terminus when either the HRV14 2A^{pro} or the HRV2 2A^{pro} cleavage site was present (Fig. 6.4). RNAs were transcribed *in vitro* and translated in RRLs. Figure 6.4 shows that the HRV14/2 2A^{pro} hybrid was unable to cleave either of the substrates. Even after 300 min of incubation, the only product produced was the uncleaved VP1-2A^{pro} running at 50 kDa; cleavage products running at 35 kDa (VP1) and 17 kDa (2A^{pro}) were absent (Fig. 6.4).

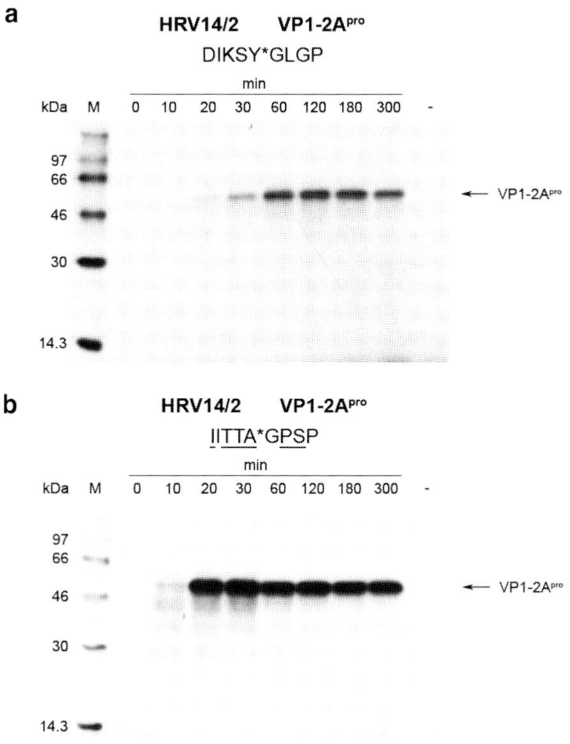

Fig. 6.4 Self-processing of HRV14/2 2A^pro^ hybrid proteinase on the HRV14 (**a**) and HRV2 (**b**) cleavage site. The amino acids in the cleavage sites are shown above the gels. Differences between the sites are underlined in the HRV2 sequence (**b**). Rabbit reticulocyte lysate was programmed in the presence of ^{35}S methionine with *in vitro* transcribed RNA coding for HRV14/2 VP1–2A^{pro} (10 ng/µl) and incubated at 30°C. Negative controls were prepared by adding water instead of RNA. 10 µl aliquots were taken at the given time points and put on an icecold mix of 25 µl 2x Laemmli sample buffer, 15 µl H_2O and 1 µl unlabeled methionine/cysteine (20 mM) Viral proteins were then separated by SDS-PAGE on 17.5% gels and visualized by fluorography. Protein standards (M) in kDa are given on the left

The hybrid 2A^{pro} is clearly inactive. What might the reasons for this be? To answer this question, we examined the nature of the residues that comprise the substrate binding region at the interface of the N- and C-terminal domains in HRV2 and HRV14 2A^{pro}. Figure 6.5 shows that only 6 of the 20 residues lining the interface of the N- and C-terminal domains are identical in the 2A^{pro} of the two genetic groups. Furthermore, three of these are involved in generating the active site triad. In contrast, this is not the case for the picornaviral 3C^{pro}; in this enzyme, the opposite is true and most of the residues lining the domain interface are conserved (Fig. 6.2). These differences in 2A^{pro} between the genetic groups appear from the experiment in Fig. 6.4 to affect substrate specificity and will have to be born in mind when designing inhibitors that should possess a broad spectrum of activity

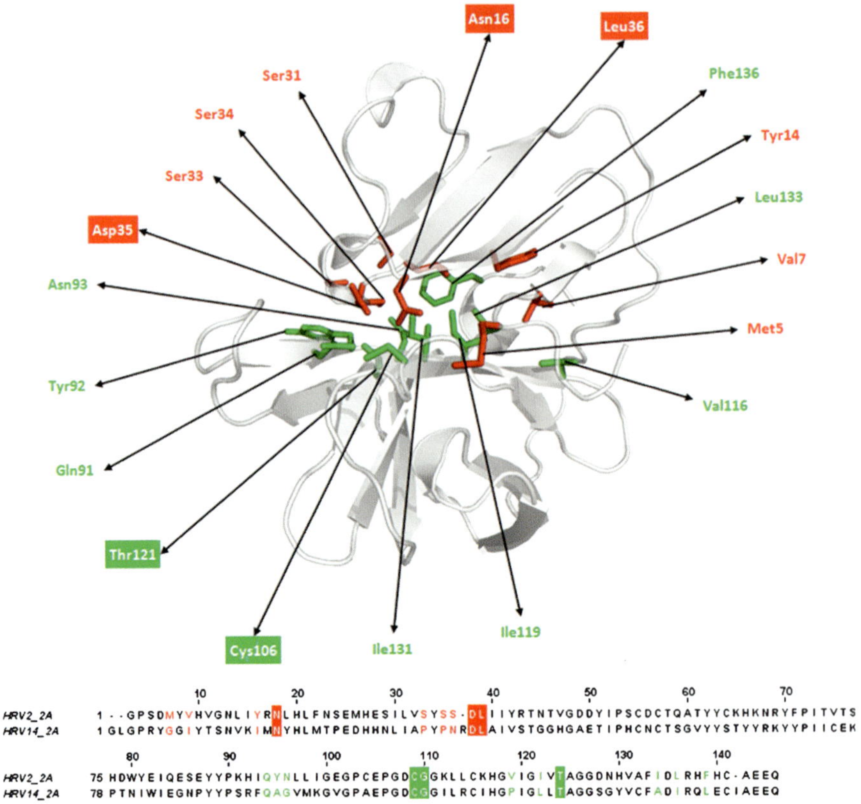

Fig. 6.5 Residues situated at the interface between the N-terminal and C-terminal domain of HRV2 2A^{pro}. (**a**) The 2A^{pro} molecule is shown as a grey ribbon at half-transparency to allow all domain interface residues to be visualised. Residues at the domain interface are shown as sticks. (**b**) Alignment of HRV2 and HRV14 2A^{pro} sequences. N-terminal domain residues at the domain interface are coloured red, C-terminal domain residues at the domain interface are in green. Reverse video in the labels and the alignment indicates conserved residues between HRV2 and HRV14 2A^{pro}. (structure generated with PyMOL DeLano, 2002) and JalView (Clamp et al., 2004)

against HRV 2A^{pro}. Furthermore, this observation again underlines the need for the determination of the structure of a 2A^{pro} from a genetic group B virus.

The section on the PV 3C^{pro} outlined the differences in structure and function between the 3C^{pro} and its precursor 3CD. For entero- and rhinoviral 2A^{pro}, there is the similar possibility that the adjacent protein on the polyprotein, 2B, can influence the properties of the 2A^{pro}. We have observed such an effect on the processing by 2A^{pro} in RRLs. Figure 6.6 shows that the presence of a single mutation, the substitution of isoleucine for asparagine at residue 94, four amino acids from the C-terminus of 2B, is capable of completely abrogating the activity of HRV14 2A^{pro}. Cleavage by wild-type 2A^{pro}-2B takes place over a period of 60–180 min, with the cleavage products VP1 (40 kDa) and 2A^{pro}-2B (28 kDa) being visible even after 20 min (Fig. 6.6a). In contrast, only the uncleaved precursor is visible in the I94N

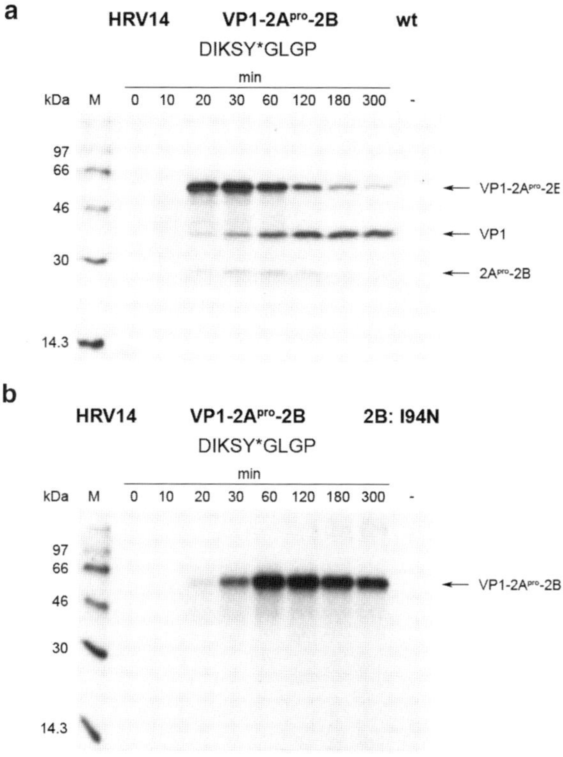

Fig. 6.6 Effect of a mutation at the C-terminus of 2B on the self-processing of HRV14 2A^{pro} on the VP1–2A^{pro}-2B polyprotein. Rabbit reticulocyte lysate was programmed with *in vitro* transcribed RNA coding for HRV14 VP1–2A^{pro}-2B wild-type (**a**) and I94N (**b**). The experiment was performed as described in the legend to Fig. 6.4

mutant (Fig. 6.6b). It will be of interest to investigate whether the differences in the inhibition profiles seen with zVAM.fmk on HRV replication may derive from differences in the properties of the 2AB precursor.

6.3.3 *Coxsackievirus 2A^{pro}*

One of the missing pieces in our understanding of the 2A^{pro} is the mechanism of self-processing. We attempted to gain information on this subject by using NMR to investigate the structure of inactive CVB4 2A^{pro} containing 8 amino acids of VP1 at the N-terminus. The enzyme was inactivated to prevent loss of the eight amino acids of VP1 by substituting the active site residue Cys110 with alanine. CVB4 2A^{pro} was used because we were unable to obtain soluble protein with the analogous HRV2 protein. We were able to assign ^{13}C and ^{15}N signals for most of the CVB4 2A^{pro} (Baxter et al., 2006). Unfortunately, however, NMR signals could not be assigned for the last C-terminal eight amino acids of VP1 and first eight N-terminal amino acids of the 2A^{pro}. The reason for this inability was determined by performing studies on the conformational dynamics of the CVB4 2A^{pro}. ^{15}N NMR relaxation measurements clearly indicated that the N-terminus of the enzyme and its VP1 extension were slowly and reversibly moving in and out of the active site, resulting in a broadening of these signals beyond detection (Baxter et al., 2006).

Structural information on the mechanism of self-processing by $2A^{pro}$ will be vital for the development of substances active against CV. We are continuing to investigate this question using X-ray crystallography on CV $2A^{pro}$ bearing a range of different VP1 extensions.

As mentioned above, the CV $2A^{pro}$ cleaves the myospecific protein dystrophin during infection (Badorff et al., 1999). This cleavage has been detected in a biopsy from a patient with dilated cardiomyopathy (Badorff and Knowlton, 2004). Encouragingly, this cleavage could be inhibited *in vitro* by a substance corresponding to the CV $2A^{pro}$ cleavage site on dystrophin (benzyloxycarbonyl-LeuSerThr-Thr–fluoromethylketone) (Badorff et al., 2000). However, the potential of this and related compounds has not yet been followed up.

6.4 Aphthoviral L Protease

6.4.1 Foot-and-Mouth Disease Virus L Protease

The Leader protease (L^{pro}) of FMDV was demonstrated to be able to free itself from the growing polypeptide chain by cleavage at its own C-terminus and the N-terminus of VP4 by Beck and Strebel (Strebel and Beck, 1986). Subsequently, Gorbalenya et al. (Gorbalenya et al., 1991) proposed that L^{pro} was a papain-like cysteine protease. This proposal was further supported by the ability of E64, a classical inhibitor of papain-like proteases, to inhibit L^{pro} (Kleina and Grubman, 1992) and modelling of the L^{pro} based on structural comparisons with those of papain and papain-like cathepsins (Skern et al., 1998). The determination of the three-dimensional structure of L^{pro} in 1998 (Guarné et al., 1998) confirmed these findings, revealing that L^{pro} has indeed a classical, if somewhat reduced, papain-like fold. In contrast to papain, however, L^{pro} has no pre-pro domain at the N-terminus; instead, it possesses a C-terminal extension (CTE) of 18 amino acids protruding from the globular domain of the enzyme. The CTE is thought to be flexible enough to reach back into the active site of the enzyme to allow intramolecular self-processing at the junction of the C-terminus with VP4. In the crystal structure, however, the CTE of one molecule was found in the active site of the adjacent one and vice-versa, thus presenting a picture of an intermolecular cleavage in the form of a homodimer.

Like the $2A^{pro}$ of rhino- and enteroviruses, L^{pro} also cleaves the eIF4G homologues, again inhibiting the translation of the host cell mRNA (Devaney et al., 1988; Gradi et al., 2004). However, attempts to find other targets for this enzyme in the infected cell have not been fruitful (Foeger, N., Hampoelz, B. and Skern, T., 2003, unpublished). Recent work has demonstrated that a portion of the L^{pro} in the infected cell translocates to the nucleus (de Los Santos et al., 2006). This raises the intriguing possibility that L^{pro} may have targets in this cellular compartment.

The inhibition of L^{pro} processing and FMDV replication shows that L^{pro} is a suitable target for anti-viral therapy (Kleina and Grubman, 1992). In addition, a recombinant FMDV lacking the L^{pro} sequence did not spread in an infected animal and appeared to be extremely sensitive to the interferon response, probably as a direct result of the absence of eIF4G cleavage (de Los Santos et al., 2006). E64, although well-tolerated by animals, is nevertheless a general inhibitor of many papain-like cysteine proteinases. It is therefore necessary to develop inhibitors that are more specific for L^{pro}.

Attempts to generate specific peptido-mimetic inhibitors of L^{pro} have however been hampered by the inability to define a consensus sequence for L^{pro} cleavage. Table 6.3 shows the cleavage sequences of the three known in vivo substrates of L^{pro}. At first glance, the sequences appear quite disparate and, on reflection, intensify the interest in the question why L^{pro} does not appear to cleave more proteins in the infected cell. However, close examination reveals that the cleavage sites imply specificity at P2 for leucine or valine and a requirement for a basic amino acid at P1, P1′ or P2′. If a basic residue is not present at one of these positions, then a small residue such as glycine or serine at P1 or P1′ or alanine or threonine at P2′ is required. Mutational analysis confirmed this interpretation, showing that L^{pro} accepted poorly substrates containing phenylalanine at P2 and did not cleave well on substrates lacking basic residues at P1, P1′ and P2′ (Glaser et al., 2001). These specificity determinants of L^{pro} are extremely unusual. It is indeed quite exceptional that a protease is able to cleave before as well as after a basic residue. This unusual property may however be advantageous in efforts to develop specific inhibitors of L^{pro}.

As with $2A^{pro}$, little is known about the mechanism of self-processing in L^{pro}. To investigate this question, we again turned to NMR, believing that this method could show how the CTE moves in and out of the active site. We also felt that measuring the structure of L^{pro} in solution would prevent the formation of the dimer observed in the crystal. We were therefore very surprised when the full-length L^{pro} behaved in solution as it did in the crystal structure by forming a homo-dimer (Cencic et al., 2007). Again, the dimer was formed by the insertion of the CTE of one molecule into the active site of the second molecule and vice versa and appeared to be further stabilised by residues at the interface of the two globular domains.

We confirmed that the C-terminus of the L^{pro} was responsible for the formation of the dimer by analysing the NMR signals of a variant of L^{pro} (termed sL^{pro}) lacking six amino acids at the C-terminus. sL^{pro} behaved as a monomer in solution, with the truncated 12 amino acid long CTE indeed showing a great degree of flexibility.

To examine the strength of dimerisation, we first added sodium chloride to the NMR sample; however, the dimer remained stable up to 2M sodium chloride, indicating that the L^{pro} dimer is a very stable one. To investigate the strength of the

Table 6.3 Known cleavage sites of L^{pro}. The arrow indicates the cleavable bond (Gradi et al., 2004; Skern et al., 2002)

Substrate	P5	P4	P3	P2	P1	↓	P1′	P2′	P3′	P4′
Polyprotein (L^{pro}–VP4)	Gln	Arg	Lys	Leu	Lys		Gly	Ala	Gly	Gln
eIF4G I	Phe	Ala	Asn	Leu	Gly		Arg	Thr	Thr	Leu
eIF4GII	Leu	Leu	Asn	Val	Gly		Ser	Arg	Arg	Ser

dimer further, we then turned to size exclusion chromatography, as this approach requires much lower amounts of protein than NMR. With this method, we were able to estimate an upper limit of 500 nM for the K_D of L^{pro} dimer dissociation. This value suggests that L^{pro} may be present as a dimer even at the low concentrations of L^{pro} found in the infected cell and thus be relevant for biological activity. This observation represents an interesting avenue of future research.

In a further attempt to examine the self-processing reaction, we decided to examine the properties of a mutant of L^{pro} in which residue leucine 200 (i.e. the last but one residue at the C-terminus corresponding to the P2 position of the substrate) had been replaced by phenylalanine. We had previously shown that this mutant was severely impaired in self-processing (Kuehnel et al., 2004; Mayer et al., 2008) and believed that the presence of the phenylalanine would impair or even completely eliminate dimer formation. This would allow us to examine the properties of a monomeric form of L^{pro} that contains a full-length CTE that could still be expected to move in and out of the active site, affecting the NMR signals of atoms in this area and providing information on self-processing.

To this end, we expressed and purified the L200F mutant of L^{pro} as described (Kirchweger et al., 1994) and examined its behaviour on size exclusion chromatography. Figure. 6.7 shows that this protein indeed behaves as a monomer, eluting at about 68 ml, the same position as sL^{pro}. In contrast, the wild-type L^{pro} elutes at about 58 ml, corresponding to the dimeric form. We are now investigating the behaviour of the CTE of the L200F mutant by NMR. Although we have not yet shown that the

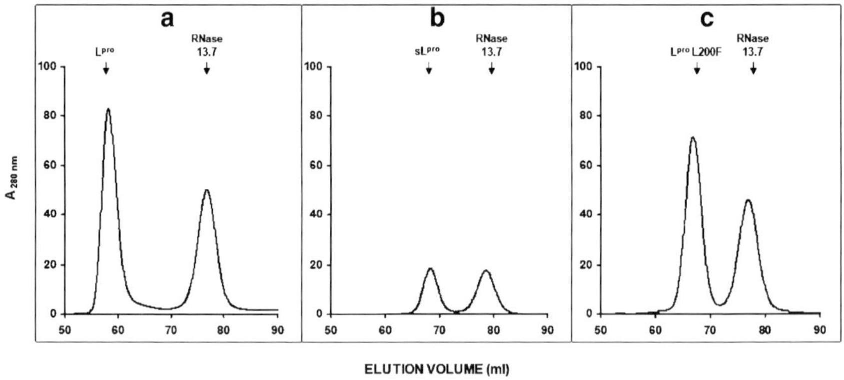

Fig. 6.7 Size exclusion chromatography of the FMDV Leader protease (L^{pro}), the shortened form (sL^{pro}) and a mutant Leader protease (L^{pro} L200F) on a HiLoad 16/60 Superdex 75 prep grade. Experiments were performed as described by Cencic et al. (Cencic et al., 2007). Ribonuclease A (RNase) with a molecular mass of 13.7 kDa was used as a reference. The elution profiles of the following combinations are shown: (**a**) L^{pro} (1 mg) and RNase (3 mg); (**b**) sL^{pro} (1 mg) and RNase (3 mg); (**c**) L^{pro} L200F (1 mg) and RNase (3 mg). Samples were applied at a flow rate of 1 ml min^{-1} in buffer A (50 mM NaCl, 50 mM Tris-HCl pH8, 1 mM EDTA, 5 mM DTT, 5% Glycerol). The apparent molecular masses in kDa are shown above the peaks. L^{pro} elutes at a volume corresponding to a molecular mass of about 40 kDa, indicating the formation of a dimer; in contrast, sL^{pro} elutes corresponding to a molecular mass of about 20 kDa, indicating a monomeric state. L^{pro} L200F elutes at a volume corresponding to a molecular mass of 20 kDa and thus resembles a monomer

dimer is present in vivo, its strength suggests that it may be involved in its biological activity. If so, the prevention of dimerisation by small molecular compounds may be another approach to interfering with the replication of FMDV.

6.4.2 *The Interaction of HRV2 2A^{pro} and L^{pro} with the Cellular Protein eIF4GI*

The above discussion concentrated on the feasibility of developing classical inhibitors of the proteolytic activity of the picornaviral proteases. In the last few years, it has however become clear that both the HRV2 2A^{pro} and the L^{pro} interact with the cellular protein eIF4G using exosites (i.e. regions of the protein away from the canonical substrate binding site). This was first observed with mutations in the CTE of L^{pro} lying ca 40 Å from the active site that interfered with the cleavage of eIF4GI without affecting the self-processing reaction (Foeger et al., 2002). Subsequently, it was shown that residues 183–195 of the CTE bound directly to the region 645–657 of eIF4GI, about 20 residues upstream of the L^{pro} cleavage site between glycine 674 and arginine 675 (Table 6.3) (Foeger et al., 2002; Foeger et al., 2005). A similar situation was also observed with HRV2 2A^{pro}. Here, residues 17–35 of the N-terminal domain, lying outside of the canonical substrate binding domain, were found to affect eIF4GI cleavage without impairing the self-processing reaction. Site-directed mutagenesis and binding experiments showed that this region of the HRV2 2A^{pro} interacted directly with a region of eIF4GI between residues 600 and 674, some distance from its actual cleavage site between amino acids arginine 681 and glycine 682 (Table 6.2) (Foeger et al., 2003). Thus, in spite of the fact that HRV2 2A^{pro} and L^{pro} are completely different enzymes in terms of structure, mechanism and specificity, both have evolved a mechanism to enable rapid cleavage of a molecule that is central to protein synthesis. At present, no host proteins have been identified as binding to eIF4G in this same manner. Therefore, inhibition of these specific interactions may also represent drug targets to fight infections with these viruses. The recent report of an inhibitor that can inhibit the interaction of eIF4GI with its binding partner eIF4E strengthens this notion (Moerke et al., 2007). The idea of inhibiting the HRV2 2A^{pro} or L^{pro} interaction with the eIF4G homologues is in line with a recent review covering the general idea of inhibiting protein-protein interactions for therapeutic purposes (Wells and McClendon, 2007). In the future, it will therefore be important to identify further specific interactions of picornaviral proteases with viral and host proteins in order to widen our options in fighting picornaviral infections.

6.5 Conclusions

Picornaviral proteinases represent attractive anti-viral targets for two reasons. First, they are absolutely required for the replication of the viruses, second, their substrate specificities differ from proteases of the host cell. Despite intensive research over the last 25 years, however, anti-virals to treat picornaviral diseases have not been approved. The most promising of all candidates, rupintrivir, unfortunately failed to reduce viral severity in the clinic and development was halted. This result implies that treatment of picornaviral infections may require two anti-viral substances attacking different viral targets for full inhibition. Although the development of a second substance as promising as rupintrivir may seem a daunting task, the work described here outlines several new avenues for the generation of such substances. Above all, the L^{pro} of FMDV and $2A^{pro}$ of rhino- and enteroviruses have not been fully examined. In addition, the possibility of inhibiting RNA binding by $3C^{pro}$ requires further investigation. All of these approaches need to be examined soon if anti-viral therapy is to play a role in the end-game of poliovirus eradication.

Acknowledgements This work was supported by a grant from the Austrian Science Foundation (P20889) to T.S.

References

Allaire, M., Chernaia, M.M, Malcolm, B.A., and James, M.N.G. 1994, Picornaviral 3C cysteine proteinases have a fold similar to chymotrypsin-like serine proteinases. *Nature* **369:** 72–76.

Almstead, L.L. and Sarnow, P. 2007, Inhibition of U snRNP assembly by a virus-encoded proteinase. *Genes Dev*. **21:** 1086–1097.

Andino, R., Rieckhof, G.E., and Baltimore, D. 1990, A functional ribonucleoprotein complex forms around the 5′ end of poliovirus RNA. *Cell* **63:** 369–380.

Andrianarivelo, M.R., Rabarijaona, L., Boisier, P., Chezzi, C., and Zeller, H.G. 1999, Wild poliovirus circulation among healthy children immunized with oral polio vaccine in antananarivo. *Madagascar Trop Med Int Health* **4:** 50–57.

Anon, 2008, Progress toward interruption of wild poliovirus transmission – Worldwide, January 2007–May 2008. *Morbidity and Mortality Weekly Report of the CDC* **57:** 489–494.

Argos, P., Kamer, G., Nickelin, M.J.H., and Wimmer, E. 1984, Similarity in gene organisation and homology between proteins of animal picornaviruses and a plant comovirus suggest a common ancestry of these virus families. *Nucl Acids Res* **12:** 7251–7267.

Armer, H., Moffat, K., Wileman, T., Belsham, G.J., Jackson, T., Duprex, W.P., Ryan, M., and Monaghan, P. 2008, Foot-and-mouth disease virus, but not bovine enterovirus, targets the host cell cytoskeleton via the nonstructural protein 3Cpro. *J Virol* **82:** 10556–10566.

Arruda, E., Pitkaranta, A., Witek, T.A., Jr., Doyle, C.A. and Hayden, F.G. 1997, Frequency and natural history of rhinovirus infections in adults during autumn. *J Clin Microbiol* **35:** 2864–2868.

Atkins, J.F., Wills, N.M., Loughran, G., Wu, C.Y., Parsawar, K., Ryan, M.D., Wang, C.H., and Nelson, C.C. 2007, A case for "StopGo": reprogramming translation to augment codon meaning of GGN by promoting unconventional termination (Stop) after addition of glycine and then allowing continued translation (Go). *RNA* **13:** 803–810.

Aylward, R.B., Sutter, R.W., and Heymann, D.L. 2005, Policy. OPV cessation – the final step to a "polio-free" world. *Science* **310***:* 625–626.

Baboonian, C., Davies, M.J., Booth, C., and McKenna, W.J. 1997, Coxsackie B viruses and human heart disease. *Curr Top Microbiol Immunol* **223:** 31–52.

Badorff, C. and Knowlton, K.U. 2004, Dystrophin disruption in enterovirus-induced myocarditis and dilated cardiomyopathy: from bench to bedside. *Med Microbiol Immunol*. **193:** 121–126.

Badorff, C., Lee, G.H., Lamphear, B.J., Martone, M.E., Campbell, K.P., Rhoads, R.E., and Knowlton, K.U. 1999, Enteroviral protease 2A cleaves dystrophin: evidence of cytoskeletal disruption in an acquired cardiomyopathy. *Nature Med* **5:** 320–326.

Badorff, C., Berkely, N., Mehrotra, S., Talhouk, J.W., Rhoads, R.E., and Knowlton, K.U. 2000, Enteroviral protease 2A directly cleaves dystrophin and is inhibited by a dystrophin-based substrate analogue. *J Biol Chem* **275:** 11191–11197.

Baxter, N.J., Roetzer, A., Liebig, H.D., Sedelnikova, S.E., Hounslow, A.M., Skern, T., and Waltho, J.P. 2006, Structure and dynamics of coxsackievirus B4 2A proteinase, an enzyme involved in the etiology of heart disease. *J Virol* **80:** 1451–1462.

Bazan, J.F. and Fletterick, R.J. 1988, Viral cysteine proteases are homologous to the trypsin-like family of serine proteases: structural and functional implications. *Proc Natl Acad Sci USA* **85:** 7872–7876.

Belsham, G.J., McInerney, G.M., and Ross-Smith, N. 2000, Foot-and-mouth disease virus 3C protease induces cleavage of translation initiation factors eIF4A and eIF4G within infected cells. *J Virol* **74:** 272–280.

Bergmann, E.M., Mosimann, S.C., Chernaia, M.M., Malcolm, B.A., and James, M.N.G. 1997, The refined crystal structure of the 3C gene product from hepatitis A virus: specific proteinase activity and RNA recognition. *J Virol* **71:** 2436–2448.

Binford, S.L., Maldonado, F., Brothers, M.A., Weady, P.T., Zalman, L.S., Meador, 3rd, J.W., Matthews, D.A., and Patick, A.K. 2005, Conservation of amino acids in human rhinovirus 3C protease correlates with broad-spectrum antiviral activity of rupintrivir, a novel human rhinovirus 3C protease inhibitor. *Antimicrob Agents Chemother* **49:** 619–626.

Birtley, J.R., Knox, S.R., Jaulent, A.M., Brick, P., Leatherbarrow, R.J., and Curry, S. 2005, Crystal structure of foot-and-mouth disease virus 3C protease. New insights into catalytic mechanism and cleavage specificity. *J Biol Chem* **280:** 11520–11527.

Bjorndahl, T.C., Andrew, L.C., Semenchenko, V., and Wishart, D.S. 2007, NMR solution structures of the apo and peptide-inhibited human rhinovirus 3C protease (Serotype 14): structural and dynamic comparison. *Biochemistry* **46:** 12945–12958.

Brundage, S.C. and Fitzpatrick, A.N. 2006, Hepatitis A. *Am Fam Physician* **73:** 2162–2168.

Cencic, R., Mayer, C., Juliano, M.A., Juliano, L., Konrat, R., Kontaxis, G., and Skern, T. 2007, Investigating the substrate specificity and oligomerisation of the leader protease of foot and mouth disease virus using NMR. *J Mol Biol.* **373:** 1071–1087.

Clamp, M., Cuff, J., Searle, S.M., and Barton, G.J. 2004, The Jalview Java alignment editor. *Bioinformatics* **20:** 426–427.

Collett, M.S., Neyts, J., and Modlin, J.F. 2008, A case for developing antiviral drugs against polio. *Antiviral Res.* **79:** 179–187.

Crowder, S., and Kirkegaard, K. 2005, Trans-dominant inhibition of RNA viral replication can slow growth of drug-resistant viruses. *Nat Genet* **37:** 701–709.

Curry, S., Roque-Rosell, N., Sweeney, T.R., Zunszain, P.A., and Leatherbarrow, R.J. 2007, Structural analysis of foot-and-mouth disease virus 3C protease: a viable target for antiviral drugs? *Biochem Soc Trans* **35:** 594–598.

DeLano, W.L. 2002, The PyMOL Molecular Graphics System, DeLano Scientific, San Carlos, CA.

de Los Santos, T., de Avila Botton, S., Weiblen, R., and Grubman, M.J. 2006, The leader proteinase of foot-and-mouth disease virus inhibits the induction of beta interferon mRNA and blocks the host innate immune response. *J Virol* **80:** 1906–1914.

De Palma, A.M., Vliegen, I., De Clercq, E., and Neyts, J. 2008, Selective inhibitors of picornavirus replication. *Med Res Rev* **28:** 823–884.

Deszcz, L., Seipelt, J., Vassilieva, E., Roetzer, A., and Kuechler, E. 2004, Antiviral activity of caspase inhibitors: effect on picornaviral 2A proteinase. *FEBS Lett* **560:** 51–55.

Deszcz, L., Cencic, R., Sousa, C., Kuechler, E., and Skern, T. 2006, An anti-viral peptide inhibitor active against picornaviral 2A proteinases but not cellular caspases. *J Virol* **80:** 9619–9627.

Devaney, M.A., Vakharia, V.N., Lloyd, R.E., Ehrenfeld, E., and Grubman, M.J. 1988, Leader protein of foot-and-mouth disease virus is required for cleavage of the p220 component of the cap-binding protein complex. *J Virol* **62:** 4407–4409.

Dragovich, P.S., Webber, S.E., Babine, R.E., Fuhrman, S.A., Patick, A.K., Matthews, D.A., Lee, C.A., Reich, S.H., Prins, T.J., Marakovits, J.T., Littlefield, E.S., Zhou, R., Tikhe, J.,

Ford, C.E., Wallace, M.B., Meador, J.W.R., Ferre, R.A., Brown, E.L., Binford, S.L., Harr, J.E., DeLisle, D.M., and Worland, S.T. 1998, Structure-based design, synthesis, and biological evaluation of irreversible human rhinovirus 3C protease inhibitors. 1. Michael acceptor structure-activity studies. *J Med Chem* **41:** 2806–2818.

Dragovich, P.S., Webber, S.E., Babine, R.E., Fuhrman, S.A., Patick, A.K., Matthews, D.A., Reich, S.H., Marakovits, J.T., Prins, T.J., Zhou, R., Tikhe, J., Littlefield, E.S., Bleckman, T.M., Wallace, M.B., Little, T.L., Ford, C.E., Meador, J.W.R., Ferre, R.A., Brown, E.L., Binford, S.L., DeLisle, D.M., and Worland, S.T. 1998, Structure-based design, synthesis, and biological evaluation of irreversible human rhinovirus 3C protease inhibitors. 2. Peptide structure-activity studies. *J Med Chem* **41:** 2819–2834.

Duechler, M, Skern, T., Sommergruber, W., Neubauer, C., Gruendler, P., Fogy, I., Blaas, D., and Kuechler, E. 1987, Evolutionary relationships within the human rhinovirus genus: comparison of serotypes 89, 2, and 14. *Proc Natl Acad Sci USA* **84:** 2605–2609.

Falk, M.M., Grigera, P.R., Bergmann, I.E., Zibert, A., Multhaup, G., and Beck, E. 1990, Foot-and-mouth disease virus protease-3C induces specific proteolytic cleavage of host cell histone-H3. *J Virol* **64:** 748–756.

Ferrer-Orta, C., Arias, A., Perez-Luque, R., Escarmis, C., Domingo, E., and Verdaguer, N. 2007, Sequential structures provide insights into the fidelity of RNA replication. *Proc Natl Acad Sci USA* **104:** 9463–9468.

Foeger, N., Glaser, W., and Skern, T. 2002, Recognition of eukaryotic initiation factor 4G isoforms by picornaviral proteinases. *J Biol Chem* **277:** 44300–44309.

Foeger, N., Schmid, E.M., and Skern, T. 2003, Human rhinovirus 2 2Apro recognition of eukaryotic initiation factor 4GI. Involvement of an exosite. *J Biol Chem* **278:** 33200–32007.

Foeger, N., Kuehnel, E., Cencic, R., and Skern, T. 2005, The binding of foot-and-mouth disease virus leader proteinase to eIF4GI involves conserved ionic interactions. *FEBS J* **272:** 2602–2611.

Glaser, W., Cencic, R., and Skern, T. 2001, Foot-and-mouth disease Leader proteinase: involvement of C-terminal residues in self-processing and cleavage of eIF4GI. *J Biol Chem* **276:** 35473–35481.

Glaser, W., Triendl, A., and Skern, T. 2003, The processing of eIF4GI by human rhinovirus 2 $2A^{pro}$: relationship to self-cleavage and role of zinc. *J Virol* **77:** 5021–5025.

Gorbalenya, A., and Svitkin, Y. 1983, Protease of encephalomyocarditis virus: purification and role of the SH groups in processing of the structural proteins precursor. *Biochemistry (USSR)* **48:** 385–395.

Gorbalenya, A.E., Donchenko, A.P., Blinov, V.M., and Koonin, E.V. 1989, Cysteine proteases of positive strand RNA viruses and chymotrypsin-like serine proteases. A distinct protein superfamily with a common structural fold. *FEBS Lett* **243:** 103–114.

Gorbalenya, A.E., Koonin, E.V., and Lai, M.M. 1991, Putative papain-related thiol proteases of positive-strand RNA viruses. Identification of rubi- and aphthovirus proteases and delineation of a novel conserved domain associated with proteases of rubi-, alpha- and coronaviruses. *FEBS Lett* **288:** 201–205.

Gouvea, I.E., Judice, W.A., Cezari, M.H., Juliano, M.A., Juhasz, T., Szeltner, Z., Polgar, L., and Juliano, L. 2006, Kosmotropic salt activation and substrate specificity of poliovirus protease 3C. *Biochemistry* **45:** 12083–12089.

Gradi, A., Svitkin, Y.V., Imataka, H., and Sonenberg, N. 1998, Proteolysis of human eukaryotic translation initiation factor eIF4GII, but not eIF4GI, coincides with the shutoff of host protein synthesis after poliovirus infection. *Proc Natl Acad Sci USA* **95:** 11089–11094.

Gradi, A., Imataka, H., Svitkin, Y.V., Rom, E., Raught, B., Morino, S., and Sonenberg, N. 1998, A novel functional human eukaryotic translation initiation factor 4G. *Mol Cell Biol* **18:** 334–342.

Gradi, A., Svitkin, Y.V., Sommergruber, W., Imataka, H., Morino, S., Skern, T., and Sonenberg, N. 2003, Human rhinovirus 2A proteinase cleavage sites in eukaryotic initiation factors (eIF) 4GI and eIF4GII are different. *J Virol* **77:** 5026–5029.

Gradi, A., Foeger, N., Strong, R., Svitkin, Y.V., Sonenberg, N., Skern, T., and Belsham, G. 2004, Cleavage of eukaryotic translation initiation factor 4GII within foot-and-mouth disease virus-infected cells: identification of the L-protease cleavage site *in vitro*. *J Virol* **78:** 3271–3278.

Graham, K.L., Gustin, K.E., Rivera, C., Kuyumcu-Martinez, N.M., Choe, S.S., Lloyd, R.E., Sarnow, P., and Utz, P.J. 2004, Proteolytic cleavage of the catalytic subunit of DNA-dependent protein kinase during poliovirus infection. *J Virol* **78:** 6313–6321.

Grubman, M.J. and Baxt, B. 2004, Foot-and-mouth disease. *Clin Microbiol Rev* **17:** 465–493.

Guarné, A., Tormo, J., Kirchweger, K., Pfistermueller, D., Fita, I., and Skern, T. 1998, Structure of the foot-and-mouth disease virus leader protease: a papain-like fold adapted for self-processing and eIF4G recognition. *EMBO J* **17:** 7469–7479.

Gustin, K.E., and Sarnow, P. 2001, Effects of poliovirus infection on nucleo-cytoplasmic trafficking and nuclear pore complex composition. *EMBO J* **20:** 240–249.

Hansen, J.L., Long, A.M., and Schultz, S.C. 1997, Structure of the RNA-dependent RNA polymerase of poliovirus. *Structure* **5:** 1109–1122.

Harki, D.A., Graci, J.D., Galarraga, J.E., Chain, W.J., Cameron, C.E., and Peterson, B.R. 2006, Synthesis and antiviral activity of 5-substituted cytidine analogues: identification of a potent inhibitor of viral RNA-dependent RNA polymerases. *J Med Chem* **49:** 6166–6169.

Hayden, F.G., Turner, R.B., Gwaltney, G.M., Chi-Burris, K., Gersten, M., Hsyu, P., Patick, A.K., Smith, 3rd, G.J., and Zalman, L.S. 2003, Phase II, randomized, double-blind, placebo-controlled studies of rupintrivir nasal spray 2-percent suspension for prevention and treatment of experimentally induced rhinovirus colds in healthy volunteers. *Antimicrob Agents Chemother.* **47:** 3907–3916.

Huitema, C., Zhang, J., Yin, J., James, M.N., Vederas, J.C., and Eltis, L.D. 2008, Heteroaromatic ester inhibitors of hepatitis A virus 3C proteinase: evaluation of mode of action. *Bioorg Med Chem* **16:** 5761–5777.

Joachims, M., Harris, K.S., and Etchison, D. 1995, Poliovirus protease 3C mediates cleavage of microtubule-associated protein 4. *Virology* **211:** 451–461.

Joachims, M., Van Breugel, P.C., and Lloyd, R.E. 1999, Cleavage of poly(A)-binding protein by enterovirus proteases concurrent with inhibition of translation *in vitro*. *J Virol* **73:** 718–727.

Johnston, S.L., Bardin, P.G., and Pattemore, P.K. 1993, Review - viruses as precipitants of asthma symptoms.3. Rhinoviruses - molecular biology and prospects for future intervention. *Clin Exp Allergy* **23:** 237–246.

Jurgens, C.K., Barton, D.J., Sharma, N., Morasco, B.J., Ogram, S.A., and Flanegan, J.B. 2006, $2A^{pro}$ is a multifunctional protein that regulates the stability, translation and replication of poliovirus RNA. *Virology* **345:** 346–357.

Katz, S.L. 2006, Polio – new challenges in 2006. *J Clin Virol.* **36:** 163–165.

Kirchweger, R., Ziegler, E., Lamphear, B.J., Waters, D., Liebig, H.D., Sommergruber, W., Sobrino, F., Hohenadl, C., Blaas, D., Rhoads, R.E., and Skern, T. 1994, Foot-and-mouth disease virus leader proteinase: purification of the Lb form and determination of its cleavage site on eIF-4 gamma. *J Virol* **68:** 5677–5684.

Kleina, L.G. and Grubman, M.J. 1992, Antiviral effects of a thiol protease inhibitor on foot-and-mouth disease virus. *J Virol* **66:** 7168–7175.

Koenig, H. and Rosenwirth, B. 1988, Purification and partial characterization of poliovirus protease 2A by means of a functional assay. *J Virol* **62:** 1243–1250.

Kuehnel, E., Cencic, R., Foeger, N., and Skern, T. 2004, Foot-and-mouth disease virus leader proteinase: specificity at the P2 and P3 positions and comparison with other papain-like enzymes. *Biochemistry* **43:** 11482–11490.

Lall, M.S., Jain, R.P., and Vederas, J.C. 2004, Inhibitors of 3C cysteine proteinases from picornaviridae. *Curr Top Med Chem.* **4:** 1239–1253.

Lau, S.K., Yip, C.C., Tsoi, H.W., Lee, R.A., So, L.Y., Lau, Y.L., Chan, K.H., Woo, P.C., and K.Y. Yuen. 2007, Clinical features and complete genome characterization of a distinct human rhinovirus (HRV) genetic cluster, probably representing a previously undetected HRV species, HRV-C, associated with acute respiratory illness in children. *J Clin Microbiol* **45:** 3655–3664.

Lee, E.S., Lee, W.G., Yun, S.H., Rho, S.H., Im, I., Yang, S.T., Sellamuthu, S., Lee, Y.J., Kwon, S.J., Park, O.K., Jeon, E.S., Park, W.J., and Kim, Y.C. 2007, Development of potent inhibitors of the coxsackievirus 3C protease. *Biochem Biophys Res Commun* **358:** 7–11.

Li, W., Ross-Smith, N., Proud, C.G., and Belsham, G.J. 2001, Cleavage of translation initiation factor 4AI (eIF4AI) but not eIF4AII by foot-and-mouth disease virus 3C protease: identification of the eIF4AI cleavage site. *FEBS Lett* **507:** 1–5.

Li, X., Lu, H.H., Mueller, S., and Wimmer, E. 2001, The C-terminal residues of poliovirus proteinase 2A(pro) are critical for viral RNA replication but not for cis- or trans-proteolytic cleavage. *J Gen Virol* **82:** 397–408.

Liebig, H.-D., Ziegler, E., Yan, R., Hartmuth, K., Klump, H., Kowalski, H., Blaas, D., Sommergruber, W., Frasel, L., Lamphear, B., Rhoads, R., Kuechler, E., and Skern, T. 1993, Purification of two picornaviral 2A proteinases: interaction with eIF-4gamma and influence on *in vitro* translation. *Biochemistry* **32:** 7581–7588.

Lloyd, R.E. 2006, Translational control by viral proteinases. *Virus Res* **119:** 76–88.

Lyle, J.M., Clewell, A., Richmond, K., Richards, O.C., Hope, D.A., Schultz, S.C., and Kirkegaard, K. 2002, Similar structural basis for membrane localization and protein priming by an RNA-dependent RNA polymerase. *J Biol Chem* **277:** 16324–16331.

MacLennan, C., Dunn, G., Huissoon, A.P., Kumararatne, D.S., Martin, J., O'Leary, P., Thompson, R.A., Osman, H., Wood, P., Minor, P., Wood, D.J., and Pillay, D. 2004, Failure to clear persistent vaccine-derived neurovirulent poliovirus infection in an immunodeficient man. *Lancet* **363:** 1509–1513.

Marcotte, L.L., Wass, A.B., Gohara, D.W., Pathak, H.B., Arnold, J.J., Filman, D.J., Cameron, C.E., and Hogle, J.M. 2007, Crystal structure of poliovirus 3CD protein: virally encoded protease and precursor to the RNA-dependent RNA polymerase. *J Virol* **81:** 3583–3596.

Matthews, D.A., Smith, W.W., Ferre, R.A., Condon, B., Budahazi, G., Sisson, W., Villafranca, J.E., Janson, C.A., McElroy, H.E., Gribskov, C.L., and Worland, S. 1994, Structure of human rhinovirus 3C protease reveals a trypsin-like polypeptide fold, RNA-binding site, and means for cleaving precursor polyprotein. *Cell* **77:** 761–771.

Matthews, D.A., Dragovich, P.S., Webber, S.E., Fuhrman, S.A., Patick, A.K., Zalman, L.S., Hendrickson, T.F., Love, R.A., Prins, T.J., Marakovits, J.T., Zhou, R., Tikhe, J., Ford, C.E., Meador, J.W., Ferre, R.A., Brown, E.L., Binford, S.L., Brothers, M.A., DeLisle, D.M., and Worland, S.T. 1999, Structure-assisted design of mechanism-based irreversible inhibitors of human rhinovirus 3C protease with potent antiviral activity against multiple rhinovirus serotypes. *Proc Natl Acad Sci USA* **96:** 11000–11007.

Mayer, C., Neubauer, D., Nchinda, A.T., Cencic, R., Trompf, K., and Skern, T. 2008, Residue L143 of the foot-and-mouth disease virus leader proteinase is a determinant of cleavage specificity. *J Virol* **82:** 4656–4659.

McErlean, P., Shackelton, L.A., Lambert, S.B. Nissen, M.D., Sloots, T.P., and Mackay, I.M. 2007, Characterisation of a newly identified human rhinovirus, HRV-QPM, discovered in infants with bronchiolitis. *J Clin Virol* **39:** 67–75.

Moerke, N.J., Aktas, H., Chen, H., Cantel, S., Reibarkh, M.Y., Fahmy, A., Gross, J.D., Degterev, A., Yuan, J., Chorev, M., Halperin, J.A., and Wagner, G. 2007, Small-molecule inhibition of the interaction between the translation initiation factors eIF4E and eIF4G. *Cell* **128:** 257–267.

Morace, G., Kusov, Y., Dzagurov, G., Beneduce, F., and Gauss-Muller, V. 2008, The unique role of domain 2A of the hepatitis A virus precursor polypeptide P1-2A in viral morphogenesis. *BMB Rep* **41:** 678–683.

Morley, S.J., Curtis, P.S., and Pain, V.M. 1997, eIF4G: translation's mystery factor begins to yield its secrets. *RNA* **3:** 1085–1104.

Mosimann, S.C., Cherney, M.M., Sia, S., Plotch, S., and James, M.N.G. 1997, Refined X-ray crystallographic structure of the poliovirus 3C gene product. *J Mol Biol* **273:** 1032–1047.

Neznanov, N., Chumakov, K.M., Neznanova, L., Almasan, A., Banerjee, A.K., and Gudkov, A.V. 2005, Proteolytic cleavage of the p65-RelA subunit of NF-kappaB during poliovirus infection. *J Biol Chem* **280:** 24153–24158.

N.R.C. Committee on Development of a Polio Antiviral and Its Potential Role in Global Poliomyelitis Eradication. 2006, Exploring the Role of Antiviral Drugs in the Eradication of Polio: Workshop Report, The National Academies Press, Washington, DC.

Ohlenschlager, O., Wohnert, J., Bucci, E., Seitz, S., Hafner, S., Ramachandran, R., Zell, R., and Gorlach, M. 2004, The structure of the stemloop D subdomain of coxsackievirus B3 cloverleaf RNA and its interaction with the proteinase 3C. *Structure* **12:** 237–248.

Park, N., Katikaneni, P., Skern, T., and Gustin, K.E. 2008, Differential targeting of nuclear pore complex proteins in poliovirus-infected cells. *J Virol* **82:** 1647–1655

Patick, A.K. 2006, Rhinovirus chemotherapy. *Antiviral Res* **71:** 391–396.

Patick, A.K., Brothers, M.A., Maldonado, F., Binford, S., Maldonado, O., Fuhrman, S., Petersen, A., Smith, 3rd, G.A., Zalman, L.S., Burns-Naas, L.A., and Tran, J.Q. 2005, *In vitro* antiviral activity and single-dose pharmacokinetics in humans of a novel, orally bioavailable inhibitor of human rhinovirus 3C protease. *Antimicrob Agents Chemother* **49:** 2267–2275.

Pelham, H.R.B. 1978, Translation of encephalomyocarditis virus RNA *in vitro* yields an active proteolytic processing enzyme. *Eur J Biochem* **85:** 457–462.

Perera, R., Daijogo, S., Walter, B.L., Nguyen, J.H., and Semler, B.L. 2007, Cellular protein modification by poliovirus: the two faces of poly(rC)-binding protein. *J Virol* **81:** 8919–8932.

Peters, H., Kusov, Y.Y., Meyer, S., Benie, A.J., Bauml, E., Wolff, M., Rademacher, C., Peters, T., and Gauss-Muller, V. 2005, Hepatitis A virus proteinase 3C binding to viral RNA: correlation with substrate binding and enzyme dimerization. *Biochem J* **385:** 363–370.

Petersen, J.F., Cherney, M.M., Liebig, H.-D., Skern, T., Kuechler, E., and James, M.N. 1999, The structure of the 2A proteinase from a common cold virus: a proteinase responsible for the shut-off of host-cell protein synthesis. *EMBO J* **18:** 5463–5475.

Racaniello, V.R. 2007, Picornaviridae: The viruses and their replication. In B.N. Fields, D.M. Knipe, and P.M. Howley (Eds.), Fields Virology, Lippincott Williams & Wilkins, Philadelphia, PA, pp. 795–838.

Renwick, N., Schweiger, B., Kapoor, V., Liu, Z., Villari, J., Bullmann, R., Miething, R., Briese, T., and Lipkin, W.I. 2007, A recently identified rhinovirus genotype is associated with severe respiratory-tract infection in children in Germany. *J Infect Dis* **196:** 1754–1760.

Robertson, S.E., Chan, C., Kim-Farley, R., and Ward, N. 1990, Worldwide status of poliomyelitis in 1986, 1987 and 1988, and plans for its global eradication by the year 2000. *World Health Stat* **Q43:** 80–90.

Sarkany, Z. and Polgar, L. 2003, The unusual catalytic triad of poliovirus protease 3C. *Biochemistry* **42:** 516–522.

Savolainen, C., Blomqvist, S., Mulders, M.N., and Hovi, T. 2002, Genetic clustering of all 102 human rhinovirus prototype strains: serotype 87 is close to human enterovirus 70. *J Gen Virol* **83:** 333–340.

Seipelt, J., Liebig, H.D., Sommergruber, W., Gerner, C., and Kuechler, E. 2000, 2A proteinase of human rhinovirus cleaves cytokeratin 8 in infected HeLa cells. *J Biol Chem* **275:** 20084–20089.

Semler, B.L. 2005, Resistance is futile. *Nat Genet* **37:** 665–666.

Semler, B.L. and Wimmer, E. 2002, Molecular Biology of Picornaviruses, ASM Press, Washington, DC

Skern, T., Sommergruber, W., Blaas, D., Gruendler, P., Frauendorfer, F., Pieler, C., Fogy, I., and Kuechler, E. 1985, Human rhinovirus 2: complete nucleotide sequence and proteolytic processing signals in the capsid protein region. *Nucl Acids Res* **13:** 2111–2126.

Skern, T., Sommergruber, W., Auer, H., Volkmann, P., Zorn, M., Liebig, H.-D., Fessl, F., Blaas, D., and Kuechler, E. 1991, Substrate requirements of a human rhinoviral 2A proteinase. *Virology* **181:** 46–54.

Skern, T., Fita, I., and Guarne, A. 1998, A structural model of picornavirus leader proteinases based on papain and bleomycin hydrolase. *J Gen Virol* **79:** 301–307.

Skern, T., Hampoelz, B., Guarné, A., Fita, I., Bergmann, E., Petersen, J., and James, M.N.G. 2002, Structure and function of picornavirus proteinases. In B.L. Semler, and E. Wimmer (Eds.), Molecular Biology of Picornaviruses, ASM Press, Washington, DC, pp. 199–212.

Sommergruber, W., Zorn, M., Blaas, D., Fessl, F., Volkmann, P., Maurer-Fogy, I., Pallai, P., Merluzzi, V., Matteo, M., Skern, T., et al. 1989, Polypeptide 2A of human rhinovirus type 2: identification as a protease and characterization by mutational analysis. *Virology* **169:** 68–77.

Sousa, C., Schmid, E.M., and Skern, T. 2006, Defining residues involved in human rhinovirus 2A proteinase substrate recognition. *FEBS Lett* **580:** 5713–5717.

Strebel, K. and Beck, E. 1986, A second protease of foot-and mouth disease virus. *J Virol* **58:** 893–899.

Svitkin, Y.V., Gradi, A., Imataka, H., Morino, S., and Sonenberg, N. 1999, Eukaryotic initiation factor 4GII (eIF4GII), but not eIF4GI, cleavage correlates with inhibition of host cell protein synthesis after human rhinovirus infection. *J Virol* **73:** 3467–3472.

Sweeney, T.R., Roque-Rosell, N., Birtley, J.R., Leatherbarrow, R.J., and Curry, S. 2007, Structural and mutagenic analysis of foot-and-mouth disease virus 3C protease reveals the role of the beta-ribbon in proteolysis. *J Virol* **81:** 115–124.

Tesar, M. and Marquardt, O. 1990, Foot-and-mouth disease virus protease 3C inhibits cellular transcription and mediates cleavage of histone H3. *Virology* **174:** 364–374.

Thomson, G.R., Vosloo, W., and Bastos, A.D. 2003, Foot and mouth disease in wildlife. *Virus Res* **91:** 145–161.

Toyoda, H., Nicklin, M.J., Murray, M.G., Anderson, C.W., Dunn, J.J., Studier, F.W., and Wimmer, E. 1986, A second virus-encoded proteinase involved in proteolytic processing of poliovirus polyprotein. *Cell* **45:** 761–770.

Turner, R.B. 2007, Rhinovirus: more than just a common cold virus. *J Infect Dis* **195:** 765–766.

Ventoso, I., MacMillan, S.E., Hershey, J.W., and Carrasco, L. 1998, Poliovirus 2A proteinase cleaves directly the eIF-4G subunit of eIF-4F complex. *FEBS Lett* **435:** 79–83.

Wells, J.C. and McClendon, C.L. 2007, Reaching for high-hanging fruit in drug discovery at protein-protein interfaces. *Nature* **450:** 1001–1009.

Witherell, G. 2000, AG-7088 Pfizer. *Curr Opin Investig Drugs* **1:** 297–302.

Yalamanchili, P., Banerjee, R., and Dasgupta, A. 1997, Poliovirus-encoded protease 2APro cleaves the TATA-binding protein but does not inhibit host cell RNA polymerase II transcription *in vitro*. *J Virol* **71:** 6881–6886.

Yin, J., Bergmann, E.M., Cherney, M.M., Lall, M.S., Jain, R.P., Vederas, J.C., and James, M.N. 2005, Dual modes of modification of hepatitis A virus 3C protease by a serine-derived beta-lactone: selective crystallization and formation of a functional catalytic triad in the active site. *J Mol Biol* **354:** 854–871.

Yoneyama, T., Yoshida, H., Shimizu, H., Yoshii, K., Nagata, N., Kew, O., and Miyamura, T. 2001, Neurovirulence of sabin 1-derived polioviruses isolated from an immunodeficient patient with prolonged viral excretion. *Dev Biol (Basel)* **105:** 93–98.

Ypma-Wong, M.F., Dewalt, P.G., Johnson, V.H., Lamb, J.G., and Semler, B.L. 1988, Protein 3CD is the major poliovirus proteinase responsible for cleavage of the P1 capsid precursor. *Virology* **166:** 265–270.

Zell, R., Sidigi, K., Bucci, E., Stelzner, A., and Gorlach, M. 2002, Determinants of the recognition of enteroviral cloverleaf RNA by coxsackievirus B3 proteinase 3C. *RNA* **8:** 188–201.

Zhang, B., Morace, G., Gauss-Muller, V., and Kusov, Y. 2007, Poly(A) binding protein, C-terminally truncated by the hepatitis A virus proteinase 3C, inhibits viral translation. *Nucleic Acids Res* **35:** 5975–5984.

Zhang, B., Seitz, S., Kusov, Y., Zell, R., and Gauss-Muller, V. 2007, RNA interaction and cleavage of poly(C)-binding protein 2 by hepatitis A virus protease. *Biochem Biophys Res Commun* **364:** 725–730.

Ziegler, E., Borman, A.M., Deliat, F.G., Liebig, H.-D., Jugovic, D., Kean, K.M., Skern, T., and Kuechler, E. 1995, Picornavirus 2A proteinase-mediated stimulation of internal initiation of translation is dependent on enzymatic activity and the cleavage products of cellular proteins. *Virology* **213:** 549–557.

Index